T0135686

Mathematical Modelling of the Jak/Stat1 Signal Transduction Pathway

Dissertation

zur Erlangung des akademischen Grades
doctor rerum naturalium
(Dr. rer. nat.)
im Fach Biophysik

eingereicht an der

Mathematisch-Naturwissenschaflichen Fakultät I
der Humboldt-Universität zu Berlin

von

Stephan Beirer
geb. 13. August 1975 in Säckingen

Tag der mündlichen Prüfung: 27. März 2007

Bibliografische Information der Deutschen Nationalbibliothek

Die Deutsche Nationalbibliothek verzeichnet diese Publikation in der Deutschen Nationalbibliografie; detaillierte bibliografische Daten sind im Internet über http://dnb.d-nb.de abrufbar.

ISBN 978-3-8325-1564-5

Logos Verlag Berlin
Comeniushof, Gubener Str. 47,
10243 Berlin
Tel.: +49 030 42 85 10 90
Fax: +49 030 42 85 10 92
INTERNET: http://www.logos-verlag.de

Contents

1

General Introduction

The ability of cells to sense changes in the environment and their capability to respond appropriately is considered a fundamental property of all biological systems - even simple life forms like bacteria react to stimuli like temperature variations or adapt to changing nutrient resources. Higher multicellular organisms have developed a complex system of intercellular communication, enabling them to coordinate the behaviour of the various cells and organs of the total organism. These signal transduction processes are critical for processing sensory information, for the coordination and regulation of cell division, and for cell differentiation and development. Often, the cellular response to an external signal entails the alteration of the activity of specific transcription factors, causing a change in gene expression that leads to the appropriate physiological response. Thus, the tight control of transcriptional regulators plays a central role in cellular communication.

During the past decades our knowledge about signal transduction networks has grown enormously: A multitude of molecular components, the specific interactions between them, and many downstream responses have been identified to date. By comparison, little is known about the quantitative aspects of signal transduction. This concerns crucial questions, as, for example: How is the transcription rate of target genes related to amplitude and duration of the external stimulus [Marshall, 1995]? In which way do the expression levels and the kinetic properties of pathway components control signalling? Which components are appropriate targets for modulating signal transduction by external interference? Answering these questions requires a quantitative understanding of signalling pathways as dynamical systems. The characterisation of the dynamics is of interest also with respect to evolutionary considerations: Which are the constraints governing the evolution of signalling systems? Can one rationalise design features as adaptations to specific objectives, such as specificity, sensitivity, and robustness?

Signal Transduction: Layered Cascades and Direct Routes into the Nucleus

A large group of transcriptional regulators are part of signalling pathways, and function by regulating the transcription of specific nuclear target genes in response to the binding of a signalling ligand to the pathway's receptor. At least two fundamental pathway designs can be differentiated. One of them is characterised by cascades of interacting proteins, with the MAP kinase group as the most prominent example. Here, consecutive interactions of kinases and phosphatases terminate at resident nuclear proteins, several of which are bound to DNA at all times. The key parameters that determine signal propagation are the activities of the participating enzymes, thus explaining how different agonists can evoke transient as well as sustained signalling of the same pathway.

The MAP kinase signalling pathway has been a prototype for mathematical modelling of cellular signal transduction systems. Starting a decade ago with a quantitative model by Huang and Ferrell [1996], the cascade has since been studied in greater detail by several groups and is probably the most modelled signalling pathway [Vayttaden et al., 2004]. With the help of these models the behaviour of MAP kinase signalling cascades could be characterised with respect to the effects of negative and positive feedback loops [Bhalla and Iyengar, 1999, Kholodenko, 2000, Asthagiri and Lauffenburger, 2001], ultra-sensitivity [Huang and Ferrell, 1996] and the appearance of bistability, hysteresis, and oscillations [Kholodenko, 2000, Ferrell and Xiong, 2001].

Direct Pathways

In contrast to cascade-like pathways in which the information is transduced via several molecular components to a nuclear transcription factor, much more direct signalling routes to the nucleus have evolved. In such systems a latent cytoplasmic transcription factor is directly activated at the cell membrane, translocates into the nucleus and switches on the expression of its target genes. Regulation of gene activation is achieved by strict control of the subcellular localisation of the transcription factor, which exists in an inactive or unstable state in the cytoplasm, while nuclear translocation is induced by activation [Komeili and O'Shea, 2000]. Examples of regulatory proteins which play a dual role as transcription factors as well as signalling molecules and where activation is controlled by nucleo-cytoplasmic translocation processes are Wnt, NFκB, Hedgehog, Notch, NFAT, Smad and the Stat factors [Darnell, 1997, Horvath, 2000, Komeili and O'Shea, 2001, Brivanlou and Darnell, 2002]. Compared with the MAP kinase-like signal transduction cascades, the mathematical modelling and the theoretical analysis of such pathways has just started recently [Lee et al., 2003, Swameye et al., 2003, Lipniacki et al.,

2004].

Jak/Stat: Cytoplasmic Activation and Nuclear Translocation

Arguably one of the simplest and most efficient signal transduction systems is the cyclic Jak/Stat pathway (**Ja**nus **K**inase/**S**ignal **T**ransducer and **A**ctivator of **T**ranscription), which is responsive to different cytokines and is involved in the immune response and in cell proliferation and differentiation. The Stat transcription factors become activated by phosphorylation on tyrosine via receptor-associated Janus kinases. Initially believed to be cytoplasmic proteins that enter the nucleus only after activation, it is now clear that the Stats are nucleo-cytoplasmic shuttling proteins both before and during the stimulation of cells with cytokines [Meyer et al., 2002b]. Best characterised in this respect is Stat1, the central mediator of the interferon response, but there appear to be many similarities in the Stat family of proteins. The unphosphorylated Stat1 molecules are capable both of carrier-dependent transport as well as carrier-independent translocation via direct interactions with the proteins of the nuclear pore. Tyrosine-phosphorylated Stat1, however, forms stable dimers that are barred from further carrier-independent transport. While nuclear import of phosphorylated Stat1 can continue in association with importin α/β, the nuclear export is precluded until the molecule is dephosphorylated. Since DNA-bound Stat1 is protected from dephosphorylation, this inactivation proceeds at a site-specific rate.

The specific activation of the Stat molecules is ensured by their SH2 domain, which recognises phosphorylated tyrosine residues in the intracellular chains of the activated receptor. The signal that emanates from the cytokine receptor is additionally modulated by numerous interactions of Jaks and Stats with regulatory proteins. Many molecular regulators that influence the activation-inactivation cycle, the DNA binding, or the recruitment of transcriptional cofactors have been characterised. In contrast, the regulatory potential that is inherent to the dynamic redistribution of the shuttling signal transducer has gained relatively little attention.

Research Objectives of This Work

Recently, several theoretical works dealing with Stat signalling have been published. Some of this theoretical approaches are not directly based on experimental data [Papin and Palsson, 2004, Zi et al., 2005]. The Stat5 model by Swameye et al. [2003] uses experimental data, but the pathway description lacks the at that time unknown shuttling process of the unphosphorylated molecules. Furthermore, the model incorporates an unphysiological delay term to account for the nuclear residence time of the activated transcription factor. In the present work, we facilitate mathematical modelling and experimental data in an integrated fashion to obtain a general picture of the design and the behaviour of the Jak/Stat1 system. We use a combination of quantitative measurements of Stat1 activation, mutagenesis, model simulations and theoretical analysis to examine the kinetic design of the Jak/Stat1 pathway and to investigate the role of the different nucleo-cytoplasmic transport processes.

In the first part of this work, a mathematical description of the Jak/Stat1 signal transduction system is developed which is able to reproduce the pathway behaviour and recovers salient features of Stat signalling. Since the behaviour of such a mathematical model is in general critically dependent on its kinetic parameters we estimate the majority of these parameters from specific experimental data and test the model accuracy by independent experimental measurements.

Using the mathematical model we characterise the pathway dynamics and seek to identify the molecular processes which regulate the Stat response. We consider the behaviour of the four major Stat pools in the system which are the concentration of phosphorylated and unphosphorylated Stat in the cytoplasm and in the nucleus, and address the following questions:

- How is the response amplitude and response duration of the Stat1 pathway determined by the rate constants of the reaction and transport steps?
- Is the control of the system response mainly determined by pathway steps upstream of the DNA binding reaction responsible for gene activation or are the remaining processes of the cycle also important?
- Do we find processes which have generally a high control over all Stat pools? Is the system robust or sensitive with respect to specific pathway processes?

The recently discovered constitutive import pathway for inactive Stat molecules appears to be a rather futile step on first sight for the function of the Jak/Stat pathway [Meyer et al., 2002b]. Since the unphosphorylated Stat molecule might have an independent function apart from activating Stat target genes, we try to determine whether this import step helps the cell to decouple the nuclear concentration of unphosphorylated Stat from the canonical cytokine response.

In the second part of this work a more global perspective is taken by considering the signal transduction process as a sequence of state transitions in a network with linear kinetics. General properties of such state transition networks are derived. Furthermore, we develop a linearised core model of Stat signalling which can be solved analytically. This simplified description of Stat signal transduction allows for analysing the system behaviour independent of specific model parameter values and enables us to approach the following more general issues:

- How is the distribution of control over the specific network processes determined by the pathway structure? Can we draw general conclusions about the magnitude of this control from the network structure alone?
- How does the import process of inactive Stat change the general behaviour of the wild-type pathway compared with a hypothetical Stat network without this import step? To what extent does this transport process change the control distribution of the system?

In the last part of this thesis we apply the results of the control analysis of part I and II directly to the biological system: The consequences of altered nucleo-cytoplasmic transport are investigated using specific Stat1 transport mutants. Based on the detailed Jak/Stat1 model established in the first part of this work we analyse the phenotypes of Stat1 mutant proteins with accelerated nuclear import and export. We consider the effects of the changed transport kinetics on the nucleo-cytoplasmic redistribution of the transcription factor and on the phosphorylation level. The results of this analysis are compared to the theoretical results of the previous sections. The effect of these mutations on the pathway kinetics and upon the transcriptional response are validated with experimental data and compared to model simulations.

This theoretical work is based on the close collaboration with the research group of Dr. UWE VINKEMEIER, Leibniz-Institut für Molekulare Pharmakalogie, 13125 Berlin. Within this group, the experimental data analysed in this work has been measured by Dr. THOMAS MEYER and colleagues.

Part I

Jak/Stat Signalling

2 The Jak/Stat1 Signal Transduction Pathway

One of the phylogenetically oldest and highly conserved signalling networks is mediated by the Janus kinases (Jak) and by signal transducer and activator of transcription (Stat) proteins. The Jak/Stat pathway is a paradigm example of signal-dependent transcription factor pathways and has been described in various species like Dictyostelium, Caenorhabditis, Anopheles, Drosophila and in mammals [Meraz et al., 1996, Darnell, 1997, Ginger et al., 2000, Horvath, 2000, Kisseleva et al., 2002]. Stat proteins are named after their dual role as signalling molecules and transcription factors. The Stat protein family consists of seven members (Stat1 - Stat4, Stat5a, Stat5b and Stat6), all of which are activated by extracellular stimuli like cytokines or growth factors. Stats are critically involved in the control of cell proliferation and differentiation and in the immune response. Dysregulation of the Stat pathway is frequently observed in cancer cells, and it has been described that Stat1- or Stat2-deficient mice are highly sensitive to microbial infections [Lau and Horvath, 2002, Levy and Darnell, 2002].

The activation of the Stat signalling system is initiated by phosphorylation of a tyrosine residue within the Stat C-terminus via receptor-associated Janus kinases. Phosphorylated Stat proteins form homo- or hetero-dimers which are imported into the nucleus where they activate their target genes [Decker et al., 1997]. After deactivation by a nuclear phosphatase, the Stats are transported back into the cytoplasm, this nuclear export allows for a cycle of continuous activation and inactivation of the signalling molecule [Andrews et al., 2002, Swameye et al., 2003].

Although initially believed to be cytoplasmic transcription factors that enter the nucleus only after activation, it is now clear that they are nucleo-cytoplasmic shuttling proteins both before and during pathway stimulation [Meyer et al.,

2002a,b, Pranada et al., 2004]. The exact biological role of this seemingly futile import and export cycle of the inactive protein is not known until now. Interestingly, such a cyclic activation-inactivation design of the pathway together with the nucleo-cytoplasmic shuttling has also been found for other transcription factors such as Smads, therefore this network topology might be a more general signalling pathway design [Inman et al., 2002, Xu and Massague, 2004].

Although there are many similarities in the Stat family members, Stat1 is best characterised with respect to the molecular events triggered by the activation of the pathway and can serve as a model system for canonical Stat signal transduction. In the following section these interactions and signalling steps in the Stat1 pathway will be described in more detail.

Overview of Stat1 Signal Transduction

Cytoplasmic Processes

Binding of the cytokine interferon-γ (IFN-γ) to its cognate receptor on the cell surface triggers a series of three tyrosine phosphorylation events (see Figure 2.1). These phosphorylations are catalyzed by the Janus kinases (Jaks), which are bound to the intracellular IFN-γ-receptor domains: Activation of the receptor by ligand binding leads to receptor dimerisation, resulting in close encounter, transphosphorylation and subsequent catalytic activation of the two Jak proteins attached to the receptor tails [Schindler et al., 1992, Greenlund et al., 1995, Heim et al., 1995, Leung et al., 1996, Boehm et al., 1997]. In turn, the activated Jaks phosphorylate a tyrosine residue within the receptor cytoplasmic domain, providing binding sites for the Src-homology-2 (SH2) domain of Stat1 proteins [Briscoe et al., 1996, Becker et al., 1998]. After binding of Stats to the intracellular receptor tail, the Jaks phosphorylate a specific tyrosine residue within the Stat proteins [Shuai et al., 1992, 1993, Schindler et al., 1995, Imada and Leonard, 2000]. The phosphorylated Stat molecules rapidly form homo-dimers by interaction of their SH2 domains [Shuai et al., 1994, Chen et al., 1998, Horvath, 2000]. There are no reports of monomeric tyrosine-phosphorylated Stat molecules, while dimer formation has been reported for inactive Stat1 and other Stats and could be a common state also for the unphosphorylated protein [Braunstein et al., 2003, Mao et al., 2005].

Transport Processes and Nuclear Interactions

Phosphorylated and activated Stat1 dimers diffuse rapidly from the plasma membrane to the nuclear pore complex (NPC) [Kerr et al., 2003]. They form a complex with the transport receptors importin α and importin β which mediate the import of cargo proteins through the NPC into the nucleus in an energy-dependent

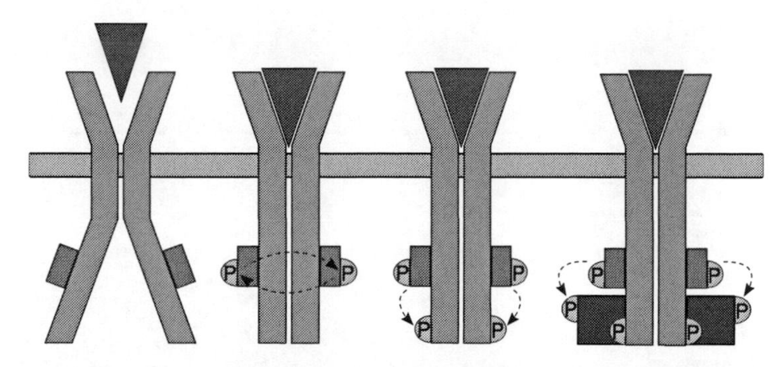

Figure 2.1: Schematic view of the activation sequence of the receptor/Jak complex and subsequent Stat phosphorylation. The ligand (red) binds to the receptor (light blue), leading to receptor dimerisation. Subsequently, the Jak kinases (orange) activate each other by transphosphorylation. The active Jaks then phosphorylate the receptor tails and two Stat proteins (dark blue) can bind to the cytoplasmic receptor domains. In the final step, the Stat molecules get phosphorylated by the active Jaks.

way. The Stats interact with the importins via a dimer-specific nuclear localisation signal (dsNLS) and get imported into the nucleus [Sekimoto et al., 1997, Fagerlund et al., 2002, McBride et al., 2002, Meyer et al., 2002a]. In the nuclear compartment, activated Stat1 dimers bind DNA unspecifically and additionally act as site-specific transcription factors by binding to their target gene recognition sequences, called GAS sites (*gamma activated sequences*) [Darnell et al., 1994, Vinkemeier et al., 1996]. Free nuclear Stat1 is deactivated rapidly by the nuclear phosphatases TC45 [David et al., 1993, Haspel et al., 1996, Haspel and Darnell, 1999, ten Hoeve et al., 2002, Wu et al., 2002], while Stat1 dimers bound both to unspecific DNA sites and to GAS sequences are effectively protected from dephosphorylation [Meyer et al., 2003, 2004]. While cytoplasmic dephosphorylation by a yet unknown phosphatase has also been observed, it is much weaker compared to the activity of the nuclear phosphatase [Tenev et al., 2000, Heinrich et al., 2003].

After deactivation, the unphosphorylated protein is exported back to the cytoplasm via two independent pathways, one employing a nuclear export signal (NES) that binds to the export receptor Crm1 [Begitt et al., 2000, Mowen and David, 2000, McBride et al., 2000]. The second pathway, which has been described recently, relies on a carrier-free interaction of Stat1 with the nuclear pore complex and acts in both directions [Meyer et al., 2002b, Marg et al., 2004]. Through

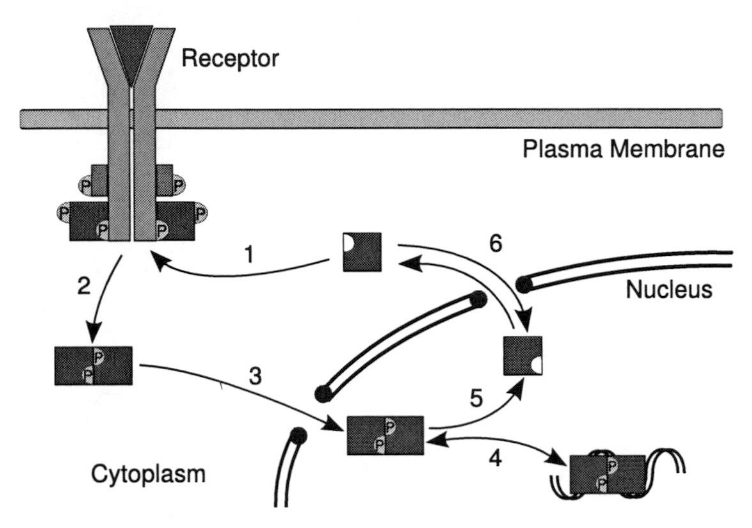

Figure 2.2: Simplified scheme of the cyclic design of the Jak/Stat1 signal transduction pathway: Unphosphorylated Stat1 binds to the active receptor/Jak complex (1), gets phosphorylated and dimerises quickly (2). The phosphorylated proteins are imported into the nucleus (3), where they bind to DNA unspecifically and to their GAS target sites (4). Free Stat becomes dephosphorylated by a nuclear phosphatase (5). Unphosphorylated Stat is exported actively back to cytoplasm and undergoes constitutive nucleo-cytoplasmic shuttling (6).

---

this energy-independent and continuous nucleo-cytoplasmic shuttling mechanism, unphosphorylated molecules can be found in the nucleus also in the absence of stimulation. Interestingly, the shuttling of the inactive protein has been found for other members of the Stat family like Stat2, Stat3 and Stat5b [Zeng et al., 2002, Banninger and Reich, 2004, Liu et al., 2005] and for other transcription factors like the Smad proteins as well [Nicolas et al., 2004, Xu and Massague, 2004].

Regulation of Stat Signalling at Multiple Levels

Numerous components and mechanisms of the molecular network regulating the Stat signal transduction have been characterised, which collectively determine the magnitude and duration of signalling [Levy and Darnell, 2002, Heinrich et al., 2003]. These include cytoplasmic and nuclear phosphatase activities [Mowen et al.,

2001, Zhu et al., 2002], tyrosine and serine phosphorylation [Decker and Kovarik, 2000], regulation of DNA binding [Meyer et al., 2003], the regulated transport processes for active and inactive Stats [Lodige et al., 2005] and the controlled turnover of the pathway components [Kim and Maniatis, 1996]. Additionally, regulation of signalling has been found by receptor internalisation and degradation as well as by tyrosine dephosphorylation of the receptor tails and Jaks, leading to progressive inhibition of the kinase activity and thus preventing further Stat activation. Negative regulators of the receptor/Jak complex are constitutive active phosphatases like SHP1, SHP2 and PTP1B [You et al., 1999, Greenhalgh and Hilton, 2001]. Furthermore, the Stats activate a negative feedback loop by inducing the transcription of SOCS proteins (Suppressor of cytokine signalling), which bind to the receptor and the Janus kinase, blocking Stat phosphorylation [Giordanetto and Kroemer, 2003]. Experimental studies have shown that the dysregulated and continuous Stat1 activity is lethal in mice [Brysha et al., 2001, Zhang et al., 2001, Metcalf et al., 2002], which underscores the importance of these regulatory mechanisms in Stat signalling. Moreover, the control of target gene expression by the duration and timing of Stat activity has been reported by Andrews et al. [2002].

3

Construction of a Mathematical Modell for Jak/Stat1 Signalling

3.1 Model of the Stat1 Pathway

Based on experimental data obtained by our collaborators U. VINKEMEIER and T. MEYER (FMP Berlin) we established a mathematical model of the Jak/Stat1 signal transduction pathway. Through an iterative process of mathematical description, computational simulations, model analysis and further experimentation to verify mechanistic assumptions and to determine kinetic parameters, the model was refined.

We describe in turn the molecular mechanisms accounted for by the Jak/Stat1 model. The pathway model is based on biological data obtained from two different experimental cell lines, U3A and HeLa cells. The two cell lines show different behaviour with respect to the receptor degradation processes. Thus, we use different parameters for this degradation mechanism in the model description of U3A and HeLa cells, while all other processes are modelled identical (details are given in section 3.2.3). The kinetic equations of the Jak/Stat1 model are formulated below using mass action kinetics and the corresponding reaction scheme is shown in Figure 3.1. For ease of notation we will denote tyrosine phosphorylated Stat1 molecules pY-Stat, while the term Y-Stat will refer to Stat1 proteins not phosphorylated at the tyrosine residue.

In our model of the Stat signalling network, the strength of an IFN-γ input stimulus is represented through the presence of a certain number of ligand-occupied receptor/Jak complexes. These receptor complexes get activated by phosphorylation (v_1), a process which is opposed by phosphatases (v_2). The IFN-bound receptor/Jak complexes are slowly degraded to account for internalisation and deactivation processes (v_3 and v_3'). Unphosphorylated Stat molecules bind to the

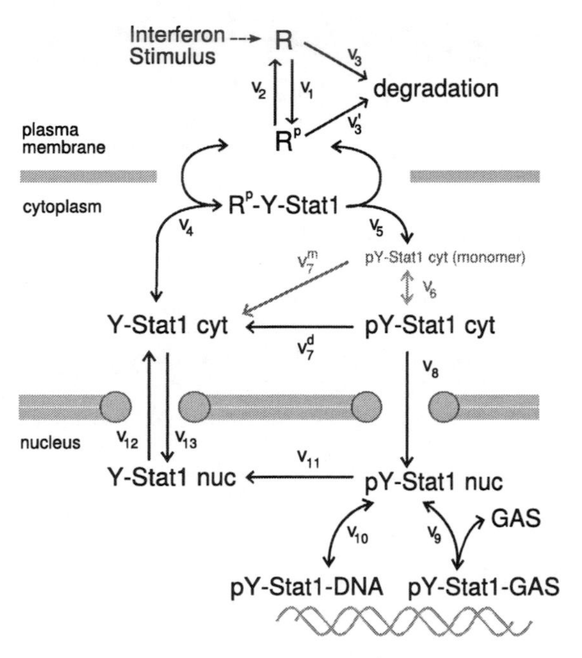

Figure 3.1: Reaction scheme of the detailed Jak/Stat1 pathway model described by equations (3.1)–(3.10). The dimerisation process has been eliminated by a quasi-steady-state approximation, thus the cytoplasmic pY-Stat monomers are not included in the mathematical model.

active receptor-Jak complexes (v_4), get phosphorylated and released (v_5), subsequently forming homo-dimers (v_6). The phosphorylated Stat monomers and dimers may become dephosphorylated by a cytoplasmic phosphatase (v_7^m and v_7^d), while the pY-Stat1 dimers are also imported into the nucleus by importins (v_8). Nuclear pY-Stat binds, both specifically (v_9) and unspecifically (v_{10}) to DNA. Free nuclear pY-Stat gets dephosphorylated (v_{11}) by a nuclear phosphatase. The unphosphorylated molecules are exported from and imported into the nucleus (v_{12} and v_{13}). The rate parameter of nuclear export includes both CRM1-dependent active export and the recently found rapid energy-independent shuttling process [Marg et al., 2004].

The concentrations of the receptor/Jak complexes and of unphosphorylated and tyrosine phosphorylated Stat molecules in the different cellular compartments are determined by the following system of balance equations:

$$\frac{d}{dt}R = -v_1 + v_2 - v_3 \qquad \text{unphos. receptor/Jak} \tag{3.1}$$

$$\frac{d}{dt}R^p = v_1 - v_2 - v_3' - v_4 + v_5 \qquad \text{phosph. receptor/Jak} \tag{3.2}$$

$$\frac{d}{dt}Y_r = v_4 - v_5 \qquad \text{receptor-bound Y-Stat1} \tag{3.3}$$

$$\frac{d}{dt}Y_c = v_{12} - v_{13} - v_4 + 2v_7 \qquad \text{cytoplasmic Y-Stat1} \tag{3.4}$$

$$\frac{d}{dt}Y_c^{p,m} = v_5 - 2v_6 - v_7^m \qquad \text{cyto. pY-Stat1 (monomer)} \tag{3.5}$$

$$\frac{d}{dt}Y_c^{p,d} = v_6 - v_7^d - v_8 \qquad \text{cyto. pY-Stat1 (dimer)} \tag{3.6}$$

$$\frac{d}{dt}Y_n^p = \rho v_8 - v_9 - v_{10} - v_{11} \qquad \text{nuclear pY-Stat1} \tag{3.7}$$

$$\frac{d}{dt}Y_n = 2v_{11} + \rho(v_{13} - v_{12}) \qquad \text{nuclear Y-Stat1} \tag{3.8}$$

$$\frac{d}{dt}Z_s = v_9 \qquad \text{bound to GAS sites} \tag{3.9}$$

$$\frac{d}{dt}Z_u = v_{10} \qquad \text{bound to unspecific DNA} \tag{3.10}$$

The rates v_i are described by mass action kinetics as given in Table 3.1 and the estimation of the model reference parameter set is described in detail in the following chapter.

The formation of Stat1 homo-dimers from tyrosine phosphorylated monomers leaving the IFN-γ/Jak complex is a very rapid, high-affinity process [Greenlund et al., 1995], thus the dimerisation process was modelled using a quasi-steady-state approximation and the concentration of pY-Stat1 monomers has been neglected in the final version of the model (see Appendix A for details of the approximation). For ease of notation we will denote the concentration of phosphorylated Stat1 in the cytoplasm with Y_c^p. The total number of Stat1 molecules in the cell is then given by

$$N_S = V_{\text{cyt}}\left(Y_c + 2\,Y_c^p + Y_r\right) + V_{\text{nuc}}\left(Y_n + 2\left(Y_n^p + Z_s + Z_u\right)\right)$$

where V_{cyt} and V_{nuc} refer to cytoplasmic and nuclear volumes, respectively, and

$$\rho = \frac{V_{\text{cyt}}}{V_{\text{nuc}}} \tag{3.11}$$

Process	Rate Equation
Receptor/Jak phosphorylation	$v_1 = k_1 R$
Receptor/Jak dephosphorylation	$v_2 = k_2 R^p$
Receptor/Jak degradation	$v_3 = k_3 R$ $v_3' = k_3 R^p$
Binding of Y-Stat1 to Receptor/Jak	$v_4 = k_4 R^p Y_c - k_{-4} Y_r$
Y-Stat1 phosphorylation	$v_5 = k_5 Y_r$
Formation of pY-Stat1 dimers	$v_6 = k_6 Y_c^{p,m} - k_{-6} Y_c^{p,d}$
Cytoplasmic dephosphorylation	$v_7^m = k_7 Y_c^{p,m}$ $v_7^d = k_7 Y_c^{p,d}$
Nuclear import of pY-Stat1	$v_8 = k_8 Y_c^p$
Binding to GAS sites	$v_9 = k_9 (G - Z_s) Y_n^p - k_{-9} Z_s$
Unspecific DNA binding	$v_{10} = k_{10} Y_n^p - k_{-10} Z_u$
Nuclear dephosphorylation	$v_{11} = k_{11} Y_n^p$
Nuclear export of Y-Stat1	$v_{12} = k_{12} Y_n$
Nuclear import of Y-Stat1	$v_{13} = k_{13} Y_c$

Table 3.1: Rate equations of the Jak/Stat1 signal transduction pathway model described by equations (3.1) - (3.10). The parameter G denotes the concentration of the specific GAS binding sites.

---

denote their ratio. Because there are no large volume differences of the nuclear and cytoplasmic compartments in the HeLa and U3A cells on which the mathematical model is based, we assume $\rho = 1$ (T. MEYER, personal communication). For the timescales investigated in this work, no change in the overall Stat concentration - for example caused by transcription or protein degradation - has been observed

[Lee et al., 1997]. Therefore, the number of Stat molecules N_S is assumed to remain constant.

The model describes a situation, where the distribution of Stat1 in the cytoplasm and in the nucleus is fast compared to reactions and transport processes across the nuclear membrane [Lillemeier et al., 2001].

In this work, the various Stat1 pools will be quantified as fractions of the total cellular Stat1 content, counted in units of Stat1 monomers. The concentration of Stat1 monomers in HeLa cells was estimated by quantitative Western blotting as 40 nM, corresponding to approximately 10^5 molecules for a spherical cell with 20 μm diameter. Thus, assuming $\rho = 1$, a Stat fraction of 1 in a given cellular compartment equals 80 nM Stat1 monomer units.

3.2 Estimation of the Stat1 Model Parameters

In the following sections, we will describe the estimation of all reference parameter values of the Jak/Stat1 pathway model. Due to difficulties in the estimation of the DNA binding parameters we will discuss this parameter estimation process in full detail and show how the difficulties were resolved by a Monte Carlo simulation approach. A summary of all model parameter reference values is given in Table 3.4 at the end of this chapter.

3.2.1 Estimation of Transport Rates

Nucleo-cytoplasmic transport rates of phosphorylated and unphosphorylated Stat1 were estimated using data from microinjection and antibody trapping experiments carried out in the lab of U. Vinkemeier. Unphosphorylated GFP-tagged Stat1 was injected in the nucleus or in the cytoplasm of HeLa and U3A cells and a pancellular steady state distribution of the fluorescent molecules was observed in less than 10 minutes for both cell types. The half-time for reaching this steady-state distribution is given by

$$t_{1/2} = \frac{\ln 2}{\rho k_{12} + k_{13}}.$$

The nucleo-cytoplasmic distribution in unstimulated cells has been measured before for various cell types in Meyer et al. [2002b]. Here we use the distribution for HeLa cells, which was estimated as 43 : 57. The nucleo-cytoplasmic ratio at steady state can be written as

$$\frac{43}{57} = \frac{V_{\text{nuc}}}{V_{\text{cyt}}}\frac{X_n}{X_c} = \rho\frac{k_{13}}{k_{12}} = 0.76$$

Using $\rho = 1$, setting the half-time to the estimated value $t_{1/2} = 2.3\,\text{min}$ and solving the two equations yields $k_{12} = 0.17\,\text{min}^{-1}$ and $k_{13} = 0.13\,\text{min}^{-1}$ for the nucleo-cytoplasmic transport rates of unphosphorylated Stat1. These rate constants also result in the correct timecourse for trapping experiments, where Stat1 antibodies are injected into the nucleus or cytoplasm of HeLa cells and the depletion of endogenous Stat1 from the opposite compartment is observed [Meyer et al., 2002b].

Microinjection of GFP-tagged phosphorylated Stat1 shows that pY-Stat1 completely accumulates in the nucleus within 5 to 10 minutes. Therefore, the import rate of pY-Stat1 was set to $k_8 = 0.3\,\text{min}^{-1}$, which yields a pY-Stat1 import half-time of 2.3 minutes. Notably, these estimates of the transport rates of unphosphorylated and tyrosine-phosphorylated Stat1 show that both transport mechanisms proceed on similar timescales.

3.2.2 Cytoplasmic and Nuclear Dephosphorylation

Dephosphorylation of Stat1 has been observed both in the nuclear compartment and in the cytoplasm. To distinguish between both dephosphorylation mechanisms, experimental data of U3A cells expressing wild-type Stat1 protein or the loss-of-function Stat1 mutant ΔNLS were analysed. The ΔNLS protein contains a mutation in the dimer specific nuclear localisation sequence (dsNLS), consequently it is not imported into the nucleus in its phosphorylated dimeric form. The mutation has been shown not to disrupt the shuttling of the unphosphorylated molecules and does not alter the ability of the Stat protein to become phosphorylated and to form dimers [Meyer et al., 2002a]. Therefore the ΔNLS mutant is phosphorylated by Jaks like the wild-type, but is protected from the nuclear phosphatase and is only dephosphorylated in the cytoplasm. When stimulated by interferon-γ, ΔNLS expressing cells show much higher and prolonged phosphorylation levels compared to the wild-type protein which indicates that the phosphatase activity in the cytoplasm is weaker than in the nucleus.

To estimate both the strength of cytoplasmic and nuclear dephosphorylation, Western blot data from mutant and wild-type cells was quantified and fitted to models describing the ΔNLS and the wild-type pathway. In the ΔNLS model, the import of phosphorylated Stat1 was set to zero ($k_8 = 0$) and the variables for phosphorylated nuclear Stat1 (Y_n^p, Z_s, Z_u) were not included in the model.

ΔNLS-Stat expressing U3A cells were incubated with the reference stimulus concentration of 5 ng/ml Interferon-γ for 45 minutes to induce maximal phosphorylation. After stimulation, the activity of the receptor bound Jak kinases was blocked by the kinase inhibitor staurosporine ($t = 0$ min), and the level of phosphorylation was measured by quantitative Western blotting. Three different data sets were available for the computational analysis: two measurements from a repetition experiment with phosphorylation values at timepoints $t = 0, 30, 90,$ and 150

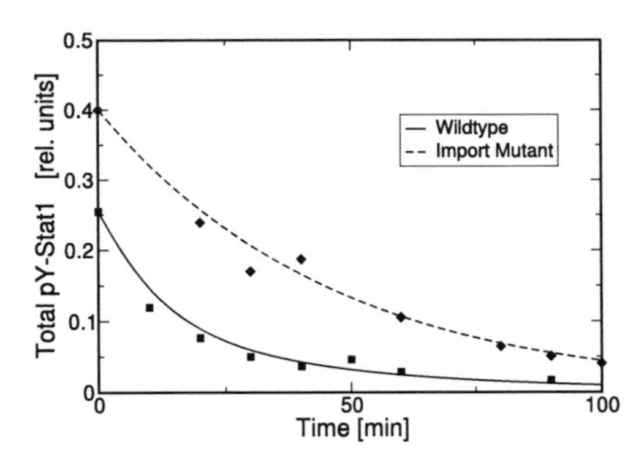

Figure 3.2: Parameter fit of a dephosphorylation assay of wild-type (straight line) and ΔNLS mutant (dashed line) Stat1. Shown are the averaged experimental data (squares and diamonds) and the obtained fits with $k_7 = 0.022\,\mathrm{min}^{-1}$ and $k_{11} = 0.09\,\mathrm{min}^{-1}$ (Experimental data: T. MEYER, FMP Berlin).

---

minutes and an additional data set with measurements at $t = 0, 20, 40, 60, 80, 100,$ and 120 minutes. All three datasets were normalised to the phosphorylation level at $t = 0$ min. Then the two datasets of the repetition experiments were averaged and merged with the third dataset, which resulted in a consistently declining time-series with small and uncorrelated fluctuations. Fitting the ΔNLS model to the dataset using MatLab's non-linear least square optimisation algorithm *lsqnonlin* (compare 3.2.5 and MathWorks [2003]) resulted in an estimated cytoplasmic first order dephosphorylation rate of $k_7 = 0.022\,\mathrm{min}^{-1}$.

To quantify the nuclear phosphatase activity, four datasets from wild-type Stat1 expressing U3A cells were analysed. Again, U3A cells were incubated with the interferon-γ reference dose of 5 ng/ml for 45 minutes to induce maximal phosphorylation, followed by inhibition of kinase activity through addition of staurosporine. The pancellular phosphorylation level was measured by quantitative Western blotting. The cell lysates were blotted at times $t = 0, 30, 90,$ and 150 minutes (datasets 1 and 2) and at $t = 0, 10, 20, 30, 40, 50, 60, 90,$ and 150 minutes (datasets 3 and 4). All four datasets were normalised to the phosphorylation level measured at $t = 0$ minutes. The datasets were averaged and fitted to the wild-type pathway model, where the cytoplasmic dephosphorylation rate was set to the previously estimated value of $k_7 = 0.022\,\mathrm{min}^{-1}$. The fitting procedure yielded an estimate for the nuclear dephosphorylation rate of $k_{11} = 0.09\,\mathrm{min}^{-1}$, which is in good agreement with a previous report by Haspel et al. [1996]. A plot of the measured dephosphorylation kinetics and the according model fits is shown in Figure 3.2.

The analysis confirmed the previous assumption that the main phosphatase activity in the Stat1 pathway is concentrated in the nucleus, yielding an approximately four times higher nuclear dephosphorylation rate. While the nuclear phosphatase accounts for the termination of the nuclear Stat1 signal, cytoplasmic phosphatases are probably responsible for preventing spontaneous and sustained activation of the pathway in the absence of extracellular stimuli.

3.2.3 Estimation of Receptor Related Parameters

The rates of the receptor/Jak complex activation reaction, Stat-receptor binding and of the subsequent Stat phosphorylation reaction were estimated using quantitative Western blot data of phosphorylated Stat1 from HeLa and U3A cells. Upon stimulation with an interferon-γ reference stimulus of 5 ng/ml, U3A cells show a rapid increase of the concentration of pY-Stat, reaching a maximum of 25-30% of the total cellular Stat content after 45 minutes. This maximum is followed by a slow decay of the pY-Stat signal. HeLa cells show a very similar onset of the phosphorylation timecourse: After stimulation the maximal amplitude of about 30-35% is reached after approximately 45 minutes as well. However, compared to U3A cells, the decline of the pY-Stat concentration is much slower in the HeLa cell line, phosphorylated Stat protein is still detectable 8 hours after stimulation (data not shown). Addition of MG132, which stabilises the phosphorylated IFN-γ-receptor/Jak complex, prolonged Stat1 phosphorylation in U3A cells significantly, indicating that the decay of the phosphorylation signal is mainly caused by internalisation and deactivation of the receptor/Jak-complex [Haspel et al., 1996]. Therefore, a first-order decay reaction (k_3) was introduced in the model to account for the loss of interferon-γ bound receptor in U3A cells.

For the reference stimulus strength the number of activateable interferon-γ receptors was set to 10000, which corresponds to 10% of the amount of Stat molecules (U. VINKEMEIER, personal communication). In the differential equation system of the mathematical model the concentrations R and R^p are expressed as relative fractions of the total Stat1 concentration of 80 nM. The rate constants of the receptor-complex phosphorylation and dephosphorylation reactions were chosen to match the onset and the maximal amplitude of the measured phosphorylation timecourse, yielding $k_1 = 0.07\,\text{min}^{-1}$ and $k_2 = 0.005\,\text{min}^{-1}$. The receptor degradation rate for U3A cells was estimated from the decay of the phosphorylation signal, yielding $k_3 = 0.017\,\text{min}^{-1}$ while for HeLa cells no receptor degradation was found on timescales up to three hours, thus we set $k_3 = 0$ for this cell type. The on- and off-rate constants for Stat1 binding to the activated IFN-γ complex have been determined in vitro by Greenlund et al. [1995] as 5×10^6 M^{-1} min^{-1}

and $k_{-4} = 0.7\ \text{min}^{-1}$, respectively. We found that an approximate doubling of the on rate constant gave quantitative agreement between model simulations and the observed phospho-Stat1 levels. Scaling by the non-dimensionalised fractional cytoplasmic receptor and Stat1 concentrations yields for $k_4 = 1\,\text{min}^{-1}$.

To determine the Stat1 phosphorylation rate individually would require a measurement of the accumulation of pY-Stat1 under specific inhibition of Stat1 phosphatases, which is currently not feasible because of the high unspecifity of available phosphatase inhibitors. Therefore, the rate constant of the Stat1 phosphorylation reaction was chosen as $k_5 = 5\ \text{min}^{-1}$ which is in agreement with the phosphorylation kinetics of other proteins [Okamura et al., 2004].

3.2.4 Specific Binding to GAS Target Sites

For Stat1 binding to its typical GAS target sites on DNA the affinities and off-rates have been determined by Vinkemeier et al. [1996] and Yang et al. [2002]. Based on these measurements we chose $K_d = 1$ nM and $k_{-9} = 0.046\,\text{min}^{-1}$, implying $k_{\text{on}} = 4.6 \times 10^7\ \text{M}^{-1}\ \text{min}^{-1}$. Scaling by the non-dimensionalised Stat reference concentration yields $k_9 = 3.68\ \text{min}^{-1}$. The number of accessible GAS sites was chosen to be $1/20$ of the number of Stat monomers, yielding a total of 5000 specific Stat1 binding sites, which is in good agreement with previous measurements on the specific binding of NFκB and Stat1 on human chromosome 22 [Martone et al., 2003, Hartman et al., 2005].

3.2.5 Estimation of Unspecific DNA Binding Parameters

Additionally to binding their primary target sequences, transcription factors also bind DNA unspecifically, which speeds up the search for their specific binding sites significantly [Halford and Marko, 2004, Kalodimos et al., 2004, Gowers et al., 2005]. To estimate a reference parameter value for the average affinity of phosphorylated Stat1 bound unspecifically to DNA, experimental data from FRAP measurements in HeLa cells was analysed. This parameter estimation demanded are more refined statistical approach, which will be described in more detail in the following sections.

FRAP Microscopy

Fluorescence Recovery after Photobleaching (FRAP) is an experimental imaging technique to visualise the mobility of macromolecules in living cells [Axelrod et al., 1976]. It allows for qualitative and quantitative determination of diffusion and binding process characteristics of the molecules under investigation. For FRAP measurements the molecules of interest are tagged with a fluorophore, for example

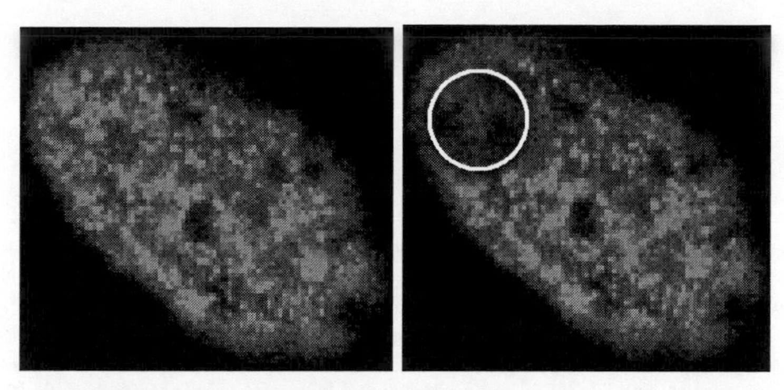

Figure 3.3: Fluorescence image of the nuclear Stat1-GFP distribution before (left) and after (right) FRAP bleaching. Shown is the nucleus of a HeLa cell after interferon-γ induced nuclear accumulation of the Stat1-GFP fusion protein. The circle approximately marks the bleached ROI (Experimental data: T. MEYER, B. WIESNER, FMP Berlin).

---

by expressing a fusion protein with *Green Fluorescent Protein* (GFP) [Miyawaki et al., 2003]. The fluorescence of the tagged molecules is irreversibly bleached by a high-powered laser in a small region of interest (ROI) and the subsequent recovery of the fluorescence signal in the ROI by diffusional redistribution of tagged molecules is monitored. Analysis of the recovery kinetics can reveal information about the tagged molecules, like diffusion constants, mobile and immobile fractions and binding/dissociation rates from other proteins [Houtsmuller and Vermeulen, 2001, Phair and Misteli, 2001, Lippincott-Schwartz et al., 2003].

For the following analysis of Stat1/DNA binding properties, previously published FRAP measurements were used, where HeLa cells expressing a Stat1-GFP fusion proteins were stimulated with IFN-γ for 30 minutes to induce nuclear accumulation of phosphorylated Stat1 dimers [Meyer et al., 2003]. 60 minutes after IFN-γ addition FRAP bleaching was performed in the nucleus with a confocal laser scanning microscope, using a 488 nm laser, a ROI diameter of 2.7 μm and a temporal resolution of 250 ms. The datasets were visually inspected and measurements with slow intensity fluctuations and oscillations (probably due to cell movements during observation time) and very high signal-to-noise ratio due to low laser intensity were discarded. Four wild-type data sets (named WT1 to WT4) and measurements of two mutant proteins were analysed. The mutants were previously characterised showing either enhanced (DNAplus) or decreased (DNAminus)

DNA-binding rates and serve as comparison to the wild-type results here [Meyer et al., 2003]. Figure 3.3 shows an example of a cell nucleus before and after bleaching of the GFP-tagged Stat1 molecules.

Modelling the Stat1 Recovery Process

The fluorescence recovery process was modelled assuming that due to the comparatively low amount of specific binding sites the majority of Stat1 dimers bind unspecifically to DNA and that the FRAP recovery kinetics is therefore determined mostly by this unspecific binding process. Testing a different model including an additional specific binding reaction did not improve the goodness-of-fit significantly. This shows that the FRAP data does not contain enough additional information about specific binding reactions. Therefore this extended model was rejected to avoid over-fitting.

Since the measured data did not provide detailed spatial information of the nucleus, a diffusion model with a simple radial symmetry was applied, where a cylindrical ROI is located in the middle of a cylindrical nucleus. The assumption of radial symmetry is justified by the fact that the cells on the microscope slide take a drop-like shape.

DNA binding was modelled as a linear first-order reaction without saturation, yielding the following reaction-diffusion system for the free (Y) and DNA-bound (Z) Stat1-GFP concentrations:

$$\begin{aligned} \dot{Y}(r,t) &= D_{\mathrm{S}} \frac{1}{r} \frac{\partial}{\partial r} \left(r \frac{\partial Y(r,t)}{\partial r} \right) - k_{\mathrm{on}} Y(r,t) + k_{\mathrm{off}} Z(r,t) && (3.12) \\ \dot{Z}(r,t) &= k_{\mathrm{on}} Y(r,t) - k_{\mathrm{off}} Z(r,t), && (3.13) \end{aligned}$$

where k_{off} is the dissociation rate from DNA, k_{on} is the first-order association rate, which includes the nuclear concentration of DNA and D_{S} denotes the nuclear diffusion coefficient of Stat1-GFP.

The diffusion coefficient D_{S} of Stat1-GFP can be roughly estimated from known coefficients of other proteins: Using the approximation that the diffusion coefficient is inverse proportional to the spatial volume of the molecule under study

$$D \propto \sqrt[-3]{M},$$

with M denoting the mass of the molecule, D_{S} can be calculated from the known nuclear diffusion constant of GFP $D_{\mathrm{G}} = 57\,\mu\mathrm{m}^2\mathrm{s}^{-1}$ [Houtsmuller et al., 1999], yielding the approximation

$$D_{\mathrm{S}} \approx 27.5\,\frac{\mu\mathrm{m}^2}{\mathrm{s}} \tag{3.14}$$

for the Stat1-GFP dimer, which is in agreement with previous experimental estimates [Lillemeier et al., 2001].

Due to varying expression characteristics of the cells used in the FRAP experiments, the single-cell data sets show different total fluorescence intensity levels. To allow for the comparison of the data sets the measurements were normalised to unity by division of the measured intensity values by the average of the 20 pre-bleach data points. Furthermore, the overall fluorescence intensity of the whole microscope image declines in a linear fashion with growing observation time due to continuous background bleaching caused by the scanning laser beam. To correct for this background bleaching, the fluorescence of a small cytoplasmic area was recorded in each cell. These background data sets were fitted by linear regression and each FRAP data set was scaled with the associated background regression. Assuming that the measured light units are directly proportional to the concentration of Stat1-GFP, the data can be interpreted as fraction of total nuclear Stat1-GFP after the normalisation and background correction was applied. To reduce the total number of parameters which have to be fitted by adjusting the model to the measured data set, the ratio of the bleached volume and the total accessible nuclear volume was estimated before data fitting from the mean of the last 100 normalised intensity values after the fluorescence had recovered to equilibrium. Figure 3.4 shows an example of two pre-processed FRAP data sets and the according model fits.

Fitting Algorithm

The partial differential equation system (3.12)–(3.13) was solved with an implicit algorithm implemented in C, using fixed parameters for the volume ratio ρ and the ROI size. The diffusion coefficient D_S, the dissociation rate k_{off} and the association rate k_{on} were fitted to the measured FRAP data sets using the MatLab implementation of the non-linear least-square quasi-Newton/Levenberg-Marquardt algorithm *lsqnonlin* [Levenberg, 1944, Marquardt, 1963]. MatLab's *lsqnonlin* function solves the minimisation problem [MathWorks, 2003]:

$$\vec{p}_{\min} = \min_{\vec{p}} \frac{1}{2} f(\vec{p}) = \min_{\vec{p}} \frac{1}{2} \|\vec{F}(\vec{p})\|^2$$

with

$$f(\vec{p}) = \|\vec{F}(\vec{p})\|^2$$

and

$$\vec{F}(\vec{p}) = \begin{bmatrix} \phi(t_1, \vec{p}) - I_1 \\ \phi(t_2, \vec{p}) - I_2 \\ \vdots \\ \phi(t_n, \vec{p}) - I_n \end{bmatrix},$$

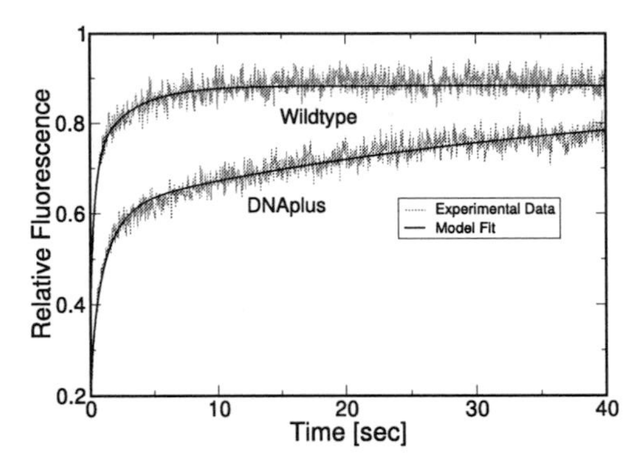

Figure 3.4: Normalised and background corrected FRAP dataset of wild-type Stat1-GFP and of the strong DNA-binding mutant DNAplus. Shown is the measured fluorescence signal and the corresponding fit. The volume ratio ρ is calculated from difference of the equilibrium intensity at the end of the timeseries (not shown for DNA-plus) to the normalisation value 1 (Experimental data: T. MEYER, B. WIESNER, FMP Berlin).

---

where $\vec{p}$ denotes the parameter vector consisting of D_S, k_{on} and k_{off}. I_i are the measured fluorescence intensities at timepoint t_i and

$$\phi(t, \vec{p}) = Y(r, t, \vec{p}) + Z(r, t, \vec{p}),\ r \in \mathrm{ROI}$$

is the sum of the modelled fluorescence intensities emitted by the free and the DNA-bound Stat1-GFP concentrations inside the ROI. Optionally the measured variance σ_i of the individual data points can be used to weight the data points according to their accuracy. However, if σ_i is constant for all datapoints, this does not influence the optimisation result. The *lsqnonlin* function returns the Jacobian matrix of $\vec{F}(\vec{p})$

$$J(\vec{p}) = \left(\frac{\partial \vec{F}_i(\vec{p})}{\partial p_j} \right)$$

evaluated at $\vec{p}_{\min}$, which can be used to gather some basic statistics about the parameter fit by calculation of the covariance matrix C. The Hessian matrix $H(\vec{p})$ of $f(\vec{p})$ can be written as [MathWorks, 2003]:

$$H(\vec{p}) = \left(\frac{\partial f(\vec{p})}{\partial p_i \partial p_j} \right) = 2J(\vec{p})^T J(\vec{p}) + 2Q(\vec{p})$$

with

$$Q(\vec{p}) = \sum_i \vec{F}_i(\vec{p}) H_i(\vec{p})$$

where $H_i(\vec{p})$ denotes the Hessian of the i-th component of $\vec{F}(\vec{p})$, $\vec{F}_i(\vec{p})$. $Q(\vec{p})$ has the property that as $\vec{p}$ approaches the solution $\vec{p}_{\text{min}}$ and $\|\vec{F}(\vec{p})\|^2$ tends to zero, then $Q(\vec{p})$ also approaches zero [MathWorks, 2003]. Therefore we can write

$$H(\vec{p}_{\text{min}}) \approx 2J(\vec{p}_{\text{min}})^T J(\vec{p}_{\text{min}}).$$

In the context of least square minimisation the curvature matrix $[\alpha]$ is defined as

$$[\alpha] = \frac{1}{2} H(\vec{p}).$$

The curvature $[\alpha]$ is the inverse of C, the estimated covariance matrix of the standard errors in the fitted parameters $\vec{p}_{\text{min}}$ [Press et al., 2002]:

$$C = [\alpha]^{-1}.$$

If $[\alpha]$ is a singular matrix the solution of the minimisation problem $\min_{\vec{p}} \frac{1}{2}\|\vec{F}(\vec{p})\|^2$ is not defined by a single point in parameter space, but by a sub-space. Then some parameters are not identifiable separately and can only be estimated in combination.

Using the covariance matrix the correlation coefficient of the parameters p_i and p_j can be calculated

$$R_{i,j} = \frac{C_{i,j}}{\sqrt{C_{i,i} C_{j,j}}}.$$

If the standard deviation σ^2 of the measurement errors is assumed to be constant for all timepoints, the elements of the covariance matrix are linear functions of σ^2 and the correlation coefficient is independent of σ^2. In this case $R_{i,j}$ can be calculated without knowledge of the strength of the noise.

Parameter Estimation

To avoid local minima the fitting procedure was applied 1000 times to each data set using random initial values for the fitted parameters. For every data set the minimisation algorithm converged to the same parameter vector in the vast majority of all trials. Other solutions were found on the border of the allowed parameter values near zero and were therefore rejected.

Because no repeated measurements under identical experimental conditions were available, the distribution of the experimental measurement errors could not

	WT1	WT2	WT3	WT4	PLUS	MINUS
D_S	0.5	1.6	1.0	0.6	0.6	0.5
k_{on}	1.54	5.24	4.12	0.70	0.83	0.62
k_{off}	6.36	20.22	9.468	8.58	1.54	4.22
$R_{D_S,k_{on}}$	0.71	0.68	0.52	0.68	0.52	0.62
$R_{D_S,k_{off}}$	0.87	0.84	0.74	0.81	0.65	0.76
$R_{k_{on},k_{off}}$	0.95	0.94	0.91	0.96	0.96	0.95

Table 3.2: Parameter estimates and correlation coefficients from initial data analysis

be estimated. Therefore it was not possible to calculate the goodness-of-fit by comparing the squared sum of residuals $f(\vec{p}_{\text{min}})$ to a χ^2-distribution. Furthermore, without knowledge of the measurement errors' standard deviation σ and the additional assumption that the errors are distributed according to a Gaussian distribution it is not possible to calculate confidence intervals on the estimated parameters from the covariance matrix C directly. This issue will be addressed in the following section.

The initial data analysis and parameter estimation produced the following results (the numerical estimates are listed in Table 3.2):

- The estimated diffusion constant D_S of all datasets is more than one order of magnitude lower than calculated in (3.14),
- estimates for k_{on} and k_{off} vary over a wide range of values and
- the correlation coefficients $R_{i,j}$ between the estimated parameters are high, especially for k_{on} and k_{off}, with $R_{k_{on},k_{off}} \approx 0.95$.

While the first observation has been reported for FRAP measurements of other proteins, too [Carrero et al., 2004], the last two points are a strong indication that the algorithm is not able to estimate both parameters k_{on} and k_{off} separately with the available datapoints, i.e. the parameters are not identifiable independently. To verify this assumption and to calculate confidence intervals for the estimated parameters without knowledge of the measurement noise characteristics, Monte Carlo simulations of Stat1 FRAP datasets were carried out.

Monte Carlo Simulation of FRAP Measurements

To calculate confidence intervals on the estimated parameters $\vec{p}_{\text{est}}$, we would like to know the distribution of all possible estimates around the true value $\vec{p}_{\text{true}}$:

$P(\vec{p}_{\text{true}} - \vec{p}_{\text{est}})$. By means of so called *Monte Carlo simulations* it is possible to determine this probability distribution without knowing $\vec{p}_{\text{true}}$ and without having access to an infinite ensemble of measured data sets and estimated parameters (for details compare Press et al. [2002]).

Let $\vec{p}_0$ denote one actual estimate of $\vec{p}_{\text{true}}$, that is one member drawn from the distribution $P(\vec{p}_{\text{est}})$ of all estimates. Now we imagine a fictitious world where $\vec{p}_{\text{true}}$ has the same value as $\vec{p}_0$ in the real world and where we have derived an ensemble of parameter estimates $\vec{p}^*_{\text{est}}$. To carry out Monte Carlo simulations, we postulate that the shape of the probability distribution $P(\vec{p}_{\text{true}} - \vec{p}_{\text{est}})$ in the real world will be identical to the shape of $P(\vec{p}_0 - \vec{p}^*_{\text{est}})$ in the fictitious world:

$$P(\vec{p}_{\text{true}} - \vec{p}_{\text{est}}) = P(\vec{p}_0 - \vec{p}^*_{\text{est}}). \tag{3.15}$$

This postulate simply expresses the fact, that the way in which random errors affect the experimental measurements and influence the data analysis should not vary intensively as a function of $\vec{p}_{\text{true}}$, that is the distribution of the estimation error should be the same around the true value $\vec{p}_{\text{true}}$ and around $\vec{p}_0$.

If the model assumptions are correct, the estimated value of $\vec{p}_0$ will be near the actual parameter value $\vec{p}_{\text{true}}$

$$\vec{p}_{\text{true}} \approx \vec{p}_0 \tag{3.16}$$

and we can use $\vec{p}_0$ as a good substitute for $\vec{p}_{\text{true}}$ when estimating $P(\vec{p}_{\text{true}} - \vec{p}_{\text{est}})$. Since $\vec{p}_0$ is known, the distribution $P(\vec{p}_0 - \vec{p}_{\text{est}})$ can be constructed if the distribution of the measurement errors is also known by simulating many artificial data sets, which are subsequently used to build an ensemble of parameter estimates [Press et al., 2002].

In the case of the used Stat1-GFP FRAP data sets, no estimate of the measurement error distribution is available, the only accessible information are the observed fluctuations in the experimental data sets. Therefore the following assumptions are made:

- Postulates (3.15) and (3.16) are valid for the Stat1-GFP experiments,
- the residual vector $\vec{F}(\vec{p}_0)$ of an initial fit $\vec{p}_0$ is an adequate estimate of the noise distribution.

After an initial estimate $\vec{p}_0$ is derived from the experimental time series $\mathcal{D}_e$, an initial dataset $\mathcal{D}_0$ is simulated with the same number of data points as $\mathcal{D}_e$, using the model (3.12)–(3.13) and $\vec{p}_0$. From $\mathcal{D}_0$, artificial Monte Carlo data sets $\mathcal{D}_i$ are generated applying a *bootstrap method* [Press et al., 2002]: The measurement

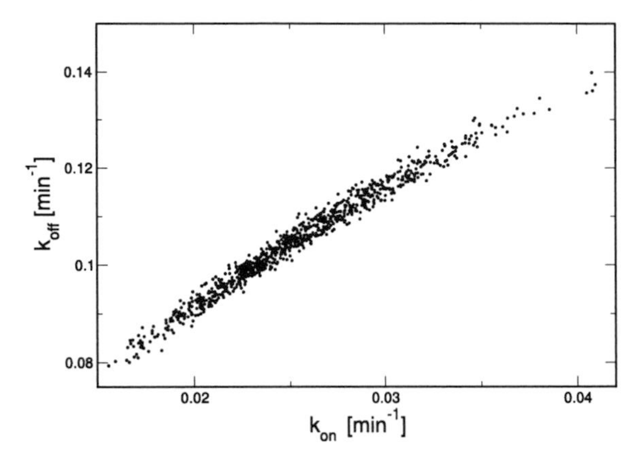

Figure 3.5: Scatter plot of the estimated Monte Carlo parameters k_{on} versus k_{off} fitted from the Stat1 FRAP data. The strong linear correlation between the estimates of the on- and off-rate constants is clearly visible.

---

noise is simulated by randomly drawing residuals from $\vec{F}(\vec{p}_0)$ and adding them to each data point in $\mathcal{D}_0$. After the generation of 1000 synthetic datasets from each of the experimental time series, the model parameters were estimated from the Monte Carlo data sets and the distribution of $P(\vec{p}_0 - \vec{p}_{\text{est}})$ was estimated by binning the estimation results. From the binned results, the most probable estimate and the smallest 68% confidence interval could be calculated.

Monte Carlo Simulation Results

An analysis of the parameter estimates generated from the Monte Carlo simulations of the wild-type data sets reveals that the binding parameters k_{on} and k_{off} are moderately correlated to the estimated diffusion constant D_{S}. A high correlation can be found for k_{on} and k_{off}, as exemplified by the scatter plot of k_{off} versus k_{on} shown in Figure 3.5. This confirms the estimated correlation coefficients of the initial fits shown in Table 3.2 and proofs that the binding parameters k_{on} and k_{off} can not be identified independently given the measured datasets.

The estimated parameters and an estimate of the association constant K_a of Stat1-GFP to DNA derived from the Monte Carlo simulations along with their 68% confidence intervals are given in Table 3.3. As in the initial fit, the diffusion coefficient is up to 2 orders of magnitude lower than estimated in equation (3.14). The estimated binding parameters and the affinity K_a vary by a factor of 4 in the

	WT1					WT2				
D_S	0.42	<	0.46	<	0.51	1.31	<	1.63	<	1.83
k_{on}	1.24	<	1.48	<	1.76	3.40	<	5.60	<	6.78
k_{off}	5.73	<	6.26	<	7.06	15.65	<	20.89	<	23.35
K_a	0.22	<	0.24	<	0.26	0.22	<	0.27	<	0.29

	WT3					WT4				
D_S	0.86	<	0.99	<	1.12	0.51	<	0.66	<	0.67
k_{on}	3.51	<	4.00	<	4.76	0.13	<	0.23	<	0.98
k_{off}	8.35	<	9.10	<	10.57	3.17	<	5.23	<	11.58
K_a	0.41	<	0.43	<	0.45	0.05	<	0.06	<	0.09

	PLUS					MINUS				
D_S	0.01	<	0.04	<	0.60	0.33	<	0.41	<	0.59
k_{on}	0.77	<	0.77	<	0.85	0.21	<	0.34	<	0.96
k_{off}	1.50	<	1.62	<	1.62	1.81	<	3.40	<	5.93
K_a	0.52	<	0.54	<	0.56	0.11	<	0.14	<	0.17

Table 3.3: Parameter estimates derived from Monte Carlo simulations. Shown are the most probable estimate and the smallest 68% confidence interval for the diffusion coefficient D_S, for the binding rates k_{on} and k_{off} and for the association constant $K_a = k_{on}/k_{off}$.

four measured wild-type data sets, which might be partly explained by biological factors, for example varying accessibility of DNA due to chromatin remodelling. The mutant proteins, which exhibit increased and decreased DNA binding show the right trend in the affinity estimates, but not for the estimates of the rate constants.

The discrepancy for the diffusion coefficient might be explained by nuclear binding processes with lower affinity on faster timescales, which are not covered by the model and which cannot be resolved with the available data sets. In this case the estimated diffusion constant D_S would be an *effective* diffusion coefficient, which incorporates the additional weakly binding process. That could also explain the wide range of binding parameters which were estimated and the observed high correlations. Similar problems have been observed with the FRAP analysis of other nuclear proteins, too [Phair and Misteli, 2000, Carrero et al., 2004]. A solution to this problem might be to use a higher temporal resolution for the initial recovery curve, which would allow to characterise the fast processes better. Additionally, better results could be found by fitting the model not only to the integrated fluorescence recovery inside the ROI, but to use the spatial fluorescence profile of the ROI as the function to fit (see [Houtsmuller et al., 1999]). Unfortunately such data was not available for the used experimental data sets.

Since the estimated association constants are in a reasonable range and show the right trend when compared to the DNAplus and DNAminus mutants, the reference binding parameters k_{10} and k_{-10} used in the Stat1 model were chosen to be similar to the observed association constants of WT1 and WT2

$$K_a = \frac{k_{10}}{k_{-10}} = \frac{1}{5} = 0.2,$$

which is in accordance with previously published experimental results published by Vinkemeier et al. [1996].

3.2.6 Parameter Reference Values

Process	Parameter Reference Values
Receptor/Jak phosphorylation	$k_1 = 0.07\ \text{min}^{-1}$
Receptor/Jak dephosphorylation	$k_2 = 0.005\ \text{min}^{-1}$
Receptor/Jak degradation	$k_3 = 0.017\ \text{min}^{-1}$ for U3A cells $k_3 = 0\ \text{min}^{-1}$ for HeLa cells
Binding of Y-Stat1 to Receptor/Jak	$k_4 = 1\ \text{min}^{-1}$ $k_{-4} = 0.7\ \text{min}^{-1}$
Y-Stat1 phosphorylation	$k_5 = 5\ \text{min}^{-1}$
Cytoplasmic dephosphorylation	$k_7 = 0.022\ \text{min}^{-1}$
Nuclear import of pY-Stat1	$k_8 = 0.3\ \text{min}^{-1}$
Binding to GAS sites	$k_9 = 3.68\ \text{min}^{-1}$ $k_{-9} = 0.046\ \text{min}^{-1}$
Unspecific DNA binding	$k_{10} = 1\ \text{min}^{-1}$ $k_{-10} = 5\ \text{min}^{-1}$
Nuclear dephosphorylation	$k_{11} = 0.09\ \text{min}^{-1}$
Nuclear export of Y-Stat1	$k_{12} = 0.17\ \text{min}^{-1}$
Nuclear import of Y-Stat1	$k_{13} = 0.13\ \text{min}^{-1}$

Table 3.4: Summary of the reference parameter values of the Jak/Stat1 signal transduction pathway model described by equations (3.1)–(3.10).

4

Jak/Stat1 Model Analysis

Having defined a set of reference parameters, the computational model of Jak/Stat1 signalling allows for a detailed analysis of the pathway kinetics and for comparison of theoretical predictions and simulations with experimentally measured data.

4.1 Model Verification by Experimental Data

Stimulation of U3A Cells

The accuracy of the model predictions were tested by comparing simulated timecourses of the concentrations of different cellular Stat1 pools with experimental data obtained under various stimulation conditions. Figure 4.1 shows the data from a stimulation experiment, where U3A cells expressing the recombinant wild-type Stat1 protein were treated with interferon-γ, and the according model simulations (solid line). The cells were activated with a continuous IFN-γ stimulus and the relative phosphorylation signal was measured at thirteen consecutive timepoints by quantitative Western blotting. The data show that the total cellular concentration of pY-Stat1 increases rapidly after stimulation, reaches a maximum of approximately 30%, and decreases slowly back to the unphosphorylated state.

In an similar experiment, U3A cells expressing the loss-of-function Stat1 mutant ΔNLS where activated with an identical interferon stimulus. The ΔNLS protein undergoes nucleo-cytoplasmic shuttling and is phosphorylated normally, but due to the lack of the dimer-specific import recognition sequence it is not imported into the nucleus after phosphorylation [Meyer et al., 2002a]. To account for this behaviour in the model, the import rate constant k_8 was set to zero. Both the simulated timeseries of the total phospho-Stat1 concentration (Figure 4.1, dashed line) and the measured experimental data show a fast rise of the phosphorylation signal, similar to the behaviour of the wild-type cells. The level of phosphoryla-

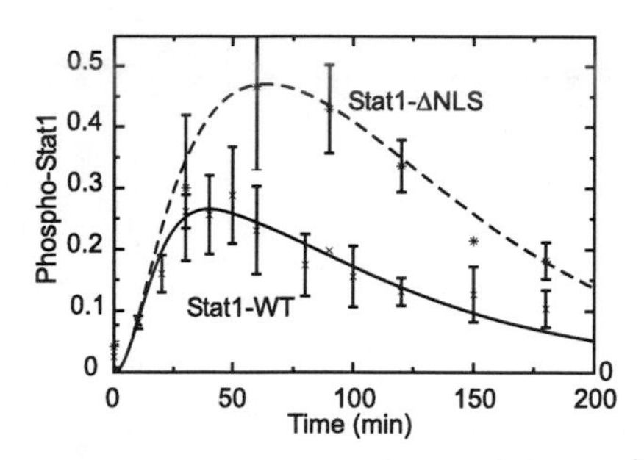

Figure 4.1: Phosphorylation level of U3A cells after interferon stimulation. Shown is the pancellular phosphorylation level of wild-type Stat1 (solid line) and the import mutant Stat1-ΔNLS (dashed line) measured by quantitative Western blotting as well as the according model simulations (Experimental data: T. MEYER, FMP Berlin).

---

tion is increased to approximately 50% and the dephosphorylation time is longer than in the wild-type system. This phosphorylation phenotype is caused by the protection of the ΔNLS mutant protein from the strong nuclear phosphatase so that the deactivation of the pathway is carried out solely by the weak cytoplasmic phosphatase.

The computational simulations of the stimulation experiments show that the model yields very good agreement with the experimentally determined data for both the ΔNLS Stat1 mutant and the wild-type protein.

HeLa Cells

Having tested the Jak/Stat1 model using U3A cells, the cell type based on which most of the model parameters were estimated, we asked how representative the computational model is for different cell types. For comparison we used experimental data from HeLa cells, where in contrast to U3A cells sustained stimulation by interferon-γ causes prolonged Stat1 phosphorylation due to reduced receptor inactivation and internalisation (cf. section 3.2.3). To reflect this difference in receptor degradation in the HeLa cell model, the decay rate of the receptor/Jak complex was set to $k_3 = 0$. Additionally, the number of receptors had to be decreased slightly to keep the maximal phosphorylation level at the experimentally determined value of $\approx 30\%$.

In an activation experiment, the HeLa cells were stimulated by IFN-γ for 45 minutes to induce a maximal phosphorylation level. Then the interferon stimulus

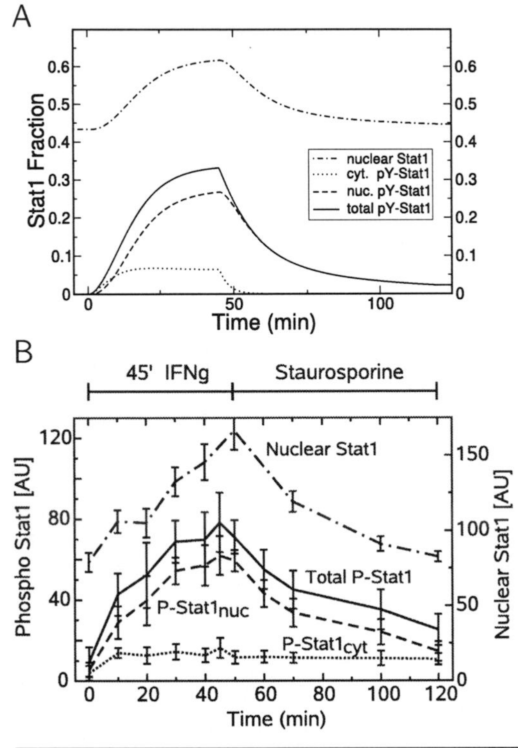

Figure 4.2: **A**: Simulated timeseries of the subcellular Stat1 fractions of a kinase inhibition experiment in HeLa cells. The addition of the kinase inhibitor staurosporine at $t = 45$ min is simulated by setting the tyrosine phosphorylation rate $k_5 = 0$.
B: Quantification of immunofluorescence data of the kinase inhibition experiment in HeLa cells. Shown is the timecourse of the nuclear accumulation of Stat1 and the cytoplasmic and nuclear phosphorylation. The total phosphorylation signal is the sum of the former two measurements (Experimental Data: T. Meyer, FMP Berlin).

was withdrawn from the cell culture and the kinase activity of the receptor/Jak complex was blocked by the inhibitor staurosporine.

The model simulations for this experimental protocol shown in Figure 4.2A predict a pancellular Stat1 distribution prior to stimulation. After stimulation by IFN-γ at $t = 0$ the simulated timeseries shows a fast increase of the cellular phosphorylation signal with most of the phosphorylated Stat molecules located in the nuclear compartment, while the cytoplasmic concentration of phospho-Stat1 stays rather small. The import of the phosphorylated protein causes the nuclear accumulation of Stat1 to increase from the steady-state value of 43% to $\approx$ 60%. Inhibition of the Jak kinase activity at the maximal phosphorylation level at $t = 45$ min causes the phosphorylation signal in the simulation to decay rapidly back to the unphosphorylated state while the nuclear-cytoplasmic distribution returns to the pancellular steady-state.

The plot in Figure 4.2B shows the quantification of immunofluorescence staining data of the stimulation experiment. Plotted are the fractions of nuclear and cytoplasmic phospho-Stat1 and the total (phosphorylated and unphosphorylated) Stat1 in the nucleus. Since the immunofluorescence data is related in a non-linear way to the actual Stat concentration, especially for small concentration levels (U. VINKEMEIER, personal communication), a direct comparison of the measured data and the computational simulation of the staurosporine experiment is not feasible. Nevertheless, the simulated time-courses of the relative nuclear Stat1 concentration and the phosphorylated fractions in the nuclear and cytoplasmic compartment in Figure 4.2A show good qualitative agreement with the experimental data. The model is able to reproduce the phosphorylation kinetics with the fast build-up and decline of pY-Stat1 and the nuclear accumulation of the transcription factor as well as the subcellular distribution with the predominant nuclear localisation of the phosphorylated protein.

When compared with the data and model simulations from U3A cells, these results demonstrate that despite the differences in the receptor dynamics of U3A and HeLa cell lines, the internal kinetics of the pathway in both cell types seem to be quite similar. Please note, while the maximal phosphorylation amplitude and the decay of the signal in the dataset shown in Figure 4.1 were used to estimate phosphorylation rate constants (cf. 3.2.3), the staurosporine experiments are independent measurements, which were not used for fitting the model parameters. Especially we had no prior knowledge of the subcellular distribution of the phosphorylated Stat1 molecules. Thus, the control experiments in HeLa and U3A cells demonstrate that the mathematical model of the Jak/Stat1 pathway is able to recover quantitatively fundamental features of the biological system in different cell types and shows that the model can be used to derive experimentally testable predictions.

4.2 Jak/Stat1 Pathway Dynamics

Strong Regulation by Phosphatase

The model simulations and the experimental data shown in Figures 4.1 and 4.2 demonstrate the rapid activation of the Stat signalling pathway after IFN-γ stimulation by the fast increase of the fraction of phosphorylated molecules. Interestingly the model predictions and the according measurements point out that Stat phosphorylation in the wild-type pathway is not maximal, only a relatively small fraction of about 30% of all Stat proteins are phosphorylated by a physiological stimulus. This low phosphorylation level indicates that phosphatases play a major role in shaping the dynamics of the pathway. Accordingly, after application of the kinase inhibitor staurosporine the system is rapidly deactivated - the HeLa cell model and simulated timeseries of the U3A wild-type system and the ΔNLS mutant proof that this inactivation is mainly driven by the strong nuclear phosphatase. The ΔNLS mutant demonstrates that the network could produce higher phosphorylation levels by limiting phosphatase activity.

The simulations of the HeLa cell staurosporine experiments (cf. Figure 4.2 A and B) suggest that the pronounced differences in the phosphorylation dynamics between the U3A and HeLa cells is not caused by differential phosphatase activities, because the pathway model is able to reproduce the experimental data with the same dephosphorylation rate parameter values for both cell types. It is sufficient to change the receptor degradation processes to explain the different phosphorylation timecourses in the two cell lines.

Furthermore, the simulated timeseries demonstrate that throughout the whole activation and deactivation process of the pathway cytoplasmic pY-Stat stays at a comparatively low concentration, while nuclear pY-Stat increases significantly. This result is consistent with the relative high rate of nuclear import (k_8) of the phosphorylated Stat dimers. The simulations show that the nuclear accumulation of Stat protein induced after pathway activation reaches a maximum of $\approx$ 60 % of all Stat molecules. Furthermore, the model demonstrates that there is no lag between phosphorylation and nuclear accumulation, likewise deactivation by dephosphorylation and the return to the pancellular steady state distribution proceed simultaneously.

Short Transition Times and Sustained Activation by Continuous Cycling

The model parameter values estimated from experimental data allow us to deduce the characteristic time constants for the transition of Stat molecules between the cytoplasmic and nuclear compartment. Under resting conditions, an unphosphorylated Y-Stat1 protein completes one cycle of shuttling from the cytoplasm to the nucleus and back in an average time of

$$\frac{1}{k_{12}} + \frac{1}{k_{13}} = 13.6\,\text{min}.$$

In the presence of a sustained interferon-γ stimulus, Stat1 will become continously phosphorylated and dephosphorylated. Then the typical residence time of a phosphorylated Stat1 dimer in the cytoplasmic compartment will be

$$\frac{1}{k_7 + k_8} = 3.1\,\text{min}.$$

Assuming a typical diffusion coefficient of 10 $\mu\text{m}^2\text{s}^{-1}$, a Stat protein diffuses several tens of μm during this time, so that diffusion from the plasma membrane to the nuclear pore is not a limiting step for Stat1 signalling in a cell with a typical diameter of 20 μm. Due to the low cytoplasmic phosphatase activity, about 93% of phosphorylated Stat1 will be imported into nucleus and be dephosphorylated there, only 7% become dephosphorylated in the cytoplasm. The small cytoplasmic dephosphorylation rate compared with the relatively high diffusion constant [Lillemeier et al., 2001] explains why there are no pronounced phosphorylation gradients as predicted for other proteins [Brown and Kholodenko, 1999, Kholodenko et al., 2000]. Therefore the cytoplasmic phosphatase does not impede effective communication from the cytoplasm to the nucleus and is expected to serve only as an inhibitor of spontaneous activation of the pathway. This results also justify the approach chosen in this work to model the Stat1 system by ordinary differential equations instead of using a reaction-diffusion model and partial differential equations.

The fast decline of the phosphorylation signal in kinase inhibitor experiments shows that under physiological conditions with continuous kinase activity the Stat molecules are re-phosphorylated several times to maintain a prolonged phosphorylation. Thus, for sustained activation of the pathway a continuous cycle of phosphorylation, nuclear import, dephosphorylation and export back to cytoplasm is essential. This result is in accordance with previous report by Swameye et al. [2003], where the continuous cycling of Stat5 was shown by mathematical modelling. The Jak/Stat1 model allows us to calculate the average cycle time for a phosphorylated Stat molecule in the presence of persistent Jak kinase activity. Under such conditions the system will attain a steady state and the steady state flux through the cycle, for example calculated by the expression $J_{SS} = k_8 Y_c^p$, will be the inverse of the average cycle time between two subsequent phosphorylation events of a Stat molecule

$$\tau_C = \frac{1}{J_{SS}}. \tag{4.1}$$

Using the reference parameter set of Table 3.4, we have for a complete cycle of phosphorylation, nuclear import, DNA binding, dephosphorylation and nuclear export of a Stat1 protein an average time of

$$\tau_C = 34\,\text{min}. \tag{4.2}$$

Small Variations of Nuclear Y-Stat1

The model is able to make detailed predictions that go beyond available experimental data by allowing us to simulate the behaviour of each subcellular Stat pool during activation and deactivation of the pathway. Figure 4.3 shows the time-dependent distribution of phosphorylated and unphosphorylated Stat1 in the nuclear and cytoplasmic compartment simulated for the IFN-γ stimulation experiment of U3A cells shown in Figure 4.1.

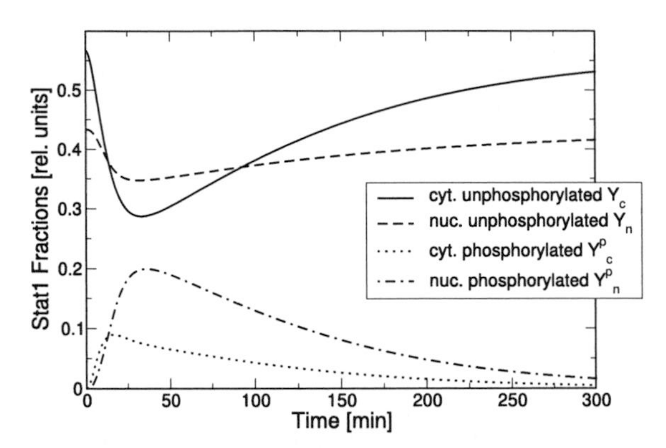

Figure 4.3: Subcellular redistribution of the main Stat1 fractions in U3A cells caused by interferon-γ stimulation. Shown is the simulated timecourse of phosphorylated and unphosphorylated Stat1 in the cytoplasm and in the nucleus (compare Figure 4.1).

The switching on of the Jak kinase decreases the concentration of the unphosphorylated cytoplasmic Stats Y_c and increases the nuclear and cytoplasmic phosphorylated fractions Y_c^p and Y_n^p. Interestingly, the pool of nuclear unphosphorylated Stat1 Y_n (dashed line) exhibits a relatively small concentration change during the activation of the pathway. Thus, the system allows for a high increase of the amount of transcriptional active phosphorylated Stat1 and relative homeostasis of unphosphorylated Stat1 in the nucleus at the same time. The model demonstrates that this property is a result of the continuous nucleo-cytoplasmic shuttling of the unphosphorylated molecules. At rest, the concentration of Y-Stat1 in the nucleus and in the cytoplasm are almost equal and Stat1 shuttles into and out of the nucleus with equal rates. As shown in Figure 4.3, stimulation of the receptor/Jak complex induces a significant drop in the concentration of unphosphorylated cytoplasmic Stat1 (Y_c, solid line), so that the net effect of shuttling becomes nuclear export. If the import rate k_{13} would be significantly lower than the reference value, then receptor activation would lead to a huge drop in Y_n, because dephosphorylated molecules would quickly leave the nucleus and become re-phosphorylated. Conversely, if the import rate would be faster than the estimated reference rate, the unphosphorylated protein would remain in the nucleus and accumulate. For an intermediate range of shuttling rates, however, the increased net export of unphosphorylated Stat1 balances the increased influx

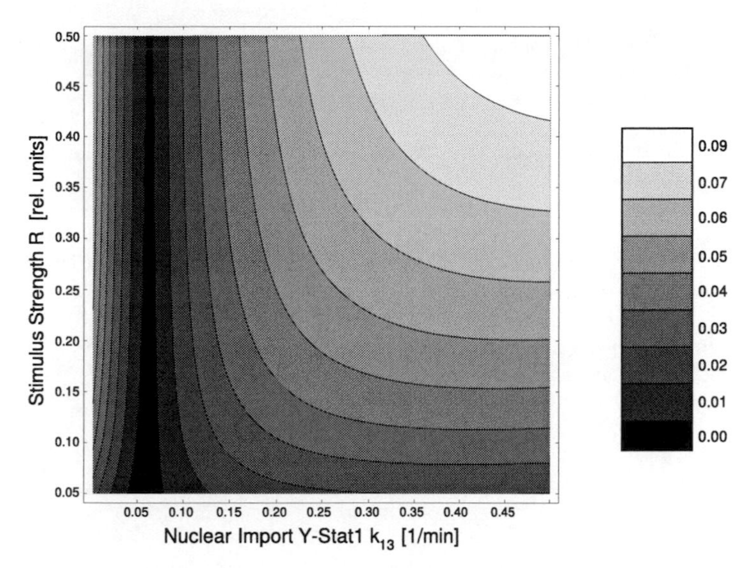

Figure 4.4: Homeostasis of nuclear Y-Stat during stimulation of the pathway. Shown is the variation σ_n of $Y_n(t)$ as a function of the import rate constant k_{12} and the stimulus strength R. The shuttling of unphosphorylated Y-Stat1 is balanced near the reference import rate $k_{13} = 0.13\,\text{min}^{-1}$, resulting in only little change of Y_n independent of the stimulus strength. Changing k_{13} results in more pronounced concentration changes of nuclear Y-Stat1 for stimulations with $R \gtrsim 0.1$.

by dephosphorylation of nuclear pY-Stat1, explaining the little change of Y_n in unstimulated and stimulated states of the pathway.

We analysed this behaviour in more detail by quantifying the magnitude of the variation of Y_n during activation of the pathway as a function of the import rate k_{13} and the stimulus strength. As a measure of the concentration change of nuclear Y-Stat1 we used the variation σ_n of $Y_n(t)$, calculated as the root of the mean squared deviation

$$\sigma_n^2 = \frac{1}{T}\int_0^T (\overline{Y}_n - Y_n(t))^2 dt.$$

The integral was taken over a total stimulation period T of five hours, after which the system has returned to the inactive state again. $\overline{Y}_n$ denotes the average of

$Y_n(t)$ over the stimulation time T. Figure 4.4 shows a contour plot of σ_n as a function of the stimulus strength and import rate k_{13}. The plot shows that over the whole range of stimulation strength, the variation of Y_n is minimal in the vicinity of the reference import rate $k_{13} = 0.13\,\text{min}^{-1}$. For higher or lower import rates the variation of Y_n increases strongly. We will show later that this behaviour can be explained in terms of the control of the receptor on Y_n, which is strongly reduced by the nucleo-cytoplasmic shuttling of the unphosphorylated Stat molecules. This analysis reveals how the balanced import and export of the inactive molecules minimises the concentration change of the unphosphorylated molecules during IFN-γ stimulation of the Jak/Stat1 pathway. Since the unphosphorylated Stat1 proteins are supposed to play an role in the transcriptional regulation of interferon independent target genes [Chatterjee-Kishore et al., 2000, Yang et al., 2005], the well-adjusted shuttling might help the cell to decouple the phosphorylated and unphosphorylated Stat fractions and their different transcriptional targets.

5

Regulation of Stat1 Activation

Individual cells show variations and heterogeneity in the expression of their molecular components - both between different cell types and among individual cells from the same family [Pedraza and van Oudenaarden, 2005, Rosenfeld et al., 2005, Mettetal et al., 2006]. One would like to know how such heterogeneity affects the behaviour and the response characteristics of cellular signalling. In the following sections we want to address this general problem with respect to the properties of the Jak/Stat signalling network and tackle the following questions:

- How robust is the pathway behaviour to variations in the biochemical parameters and which processes are critical for achieving the nuclear accumulation of the active transcription factor?
- How do intrinsic kinetic properties of the Jak/Stat network and extrinsic stimulus properties shape the cellular response to cytokine signals?
- To what extend can the pathway response be regulated by the expression levels and activities of the molecular components of the signal transduction system?

5.1 Response Control Analysis

Often, signal transduction networks transform an external stimulus into a change of the expression pattern of one or several target genes. In Jak/Stat signalling, the concentration of phosphorylated Stat1 in the nucleus determines the probability of the occupancy of GAS target sites and will hence be critical for target gene transcription [Levy and Darnell, 2002]. Furthermore, the duration of the nuclear pY-Stat1 signal will affect the expression pattern of the Stat target genes [Heinrich et al., 2003].

Therefore we computed how changes in the kinetic parameters affect the nuclear phospho-Stat1 concentration and signal duration. We changed the model (3.1)–(3.10) by neglecting receptor phosphorylation, dephosphorylation and degradation processes (v_1, v_2, v_3 and v_3'). Instead we applied a time-dependent stimulus, consisting of an instantaneous initial increase of the active IFN-receptor/Jak complexes R^p to a maximal value R_0, followed by a first-order decay with lifetime T_S, caused by the inactivation and degradation of the receptor/Jak complex:

$$R^p(t) = \begin{cases} 0 & \text{for } t < 0 \\ R_0 e^{-t/T_S} & \text{for } t \geq 0 \end{cases} \tag{5.1}$$

Modelling the cytokine stimulus with this simple input function allows us to analyse different modes of stimulation by varying the stimulus strength R_0 and its half-life $T_{0.5} = T_S \times \ln 2$.

The resulting transient pY-Stat1 signal in the nucleus was characterised by its amplitude α and the response duration, taken to be twice the signal's mean variation $\sigma_{p,n}$, which is defined in the following way:

$$\sigma_{p,n}^2 = \frac{1}{I_p} \int_0^\infty t^2 P_n(t) dt - \tau_{p,n}^2$$

with

$$I_p = \int_0^\infty P_n(t) dt \quad \text{and} \quad \tau_{p,n} = \frac{1}{I_p} \int_0^\infty t P_n(t) dt$$

and where $P_n(t) = Y_n^p(t) + Z_s(t) + Z_u(t)$ denotes the total fraction of phosphorylated Stat1 molecules in the nucleus. The integrated response I_p denotes the total amount of nuclear phospho-Stat1 produced throughout the stimulation period and $\tau_{p,n}$ is the average residence time of a phosphorylated Stat molecule in the nucleus (compare Figure 5.1).

The pathway parameter set given in Table 3.4 and a stimulus with a certain R_0 and $T_{0.5}$ define a reference parameter set, starting from which every parameter was varied over a range between 25% and 400% of its reference value. To compare the control exerted by the individual steps, the amplitude and response duration of nuclear pY-Stat1 were plotted as a function of the variation of each parameter. To allow for better comparison of the different processes, the parameter changes were normalised with respect to their reference values. In particular, we varied both the intrinsic parameters of the pathway model and the characteristics of the external stimulus, that is the stimulus amplitude R_0 and its duration $T_{0.5}$.

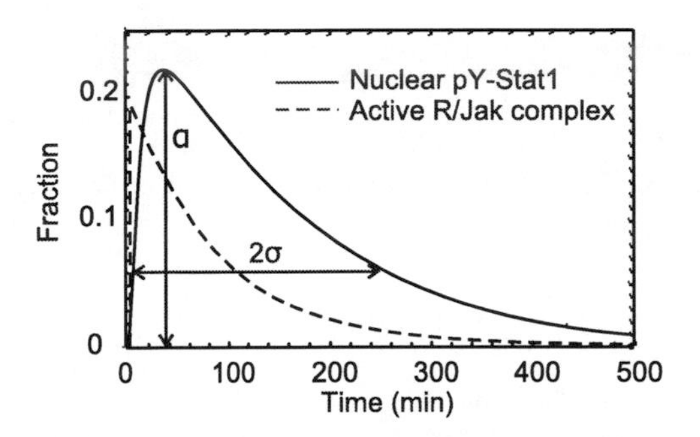

Figure 5.1: Characterisation of the transient pathway response. The stimulus activating the network is represented by an exponential decaying amount of active receptor/Jak complexes (dashed line). The resulting nuclear phosphorylation level (solid line) is analysed with respect to response amplitude α and response duration 2σ.

5.1.1 Control over Response Amplitude

Weak Stimulation

The control of the response amplitude depends on the strength of stimulation in the reference state. A weak stimulus (Figure 5.2 A, $R_0 = 0.01$ and $T_{0.5} = 60\,\text{min}$) will lead to the phosphorylation of only a small percentage of the Stat1 pool, e.g. $\approx 5\%$ at the reference point where all the curves intersect. Accordingly, an increase in nuclear phospho-Stat1 can efficiently be induced by a further increase of the stimulus through the activation of more receptor/Jak complexes (R_0, blue line). An increase in the on-rate constant k_4 of Stat1 binding to receptor/Jak complex has a similarly high control as the number of active receptors (light blue line). This finding shows that for low stimulus strengths the recruitment of Stat1 to the receptor is the limiting step for the activation of the molecules, while the phosphorylation reaction itself has a rather low control (not shown). The signal amplitude can be enhanced with nearly the same efficiency by decreasing the nuclear phosphatase activity (k_{11}, black line). Interestingly, the amplitude is also increased when more specific DNA binding sites become available (G, red line). This is the case because DNA binding of Stat1 competes with phosphatase binding and, therefore, effectively lowers the dephosphorylation rate. The next parameters in the control hierarchy are the export and import rate constants of unphosphorylated Stat1 (k_{12} and k_{13}, dark and light green lines, respectively). The remaining parameters all have much smaller effects (not shown).

Strong Stimulation

When the pathway is stimulated strongly (Figure 5.2 B, $R_0 = 1.0$ and $T_{0.5} = 60\,\text{min}$), the nuclear phosphatase retains its high negative control (k_{11}, black line). By contrast, the response amplitude of phosphorylated nuclear Stat1 has become saturated with respect to Jak kinase activity, so that a further increase of the number of active receptor/Jak complexes has little effect (blue line). The number of GAS DNA binding sites also has only little effect under this stimulation conditions (G, red line). The most significant positive control in this stimulation regime is exerted by the nuclear export k_{12} (dark green line), which under weak stimulation conditions had only a moderate impact. The import of the unphosphorylated molecule also has a noticeable negative effect on the amplitude of nuclear pY-Stat1 (k_{13}, light green line). All other parameters have a smaller, negligible control (not shown).

5.1.2 Control over Response Duration

Short Stimulation

The duration of the nuclear residence of phosphorylated Stat1 is controlled by (i) how well the protein can be retained on DNA and (ii) how long the pathway is stimulated (Figure 5.2 C and D). The weighting of these two factors depends on the halflife $T_{0.5}$ of the reference stimulus. When the reference stimulus is very short (Figure 5.2 C, $T_{0.5} = 10\,\text{min}$), several DNA binding parameters (number of GAS sites G, red line; on and off rate constants to GAS sites k_9 and k_{-9}, light and dark green lines, respectively), have strong control over the nuclear residence time of the transcriptionally active phospho-Stat1 dimers. Similarly, the nuclear phosphatase has a large negative duration control (black line). By contrast, a moderate change in the stimulus length has little effect on the duration of the response (purple line), especially stimulus durations of $T_{0.5} \lessapprox 15\,\text{min}$ have practically no impact. This result shows that there is a kinetic constraint on monitoring the receptor activity by the Stat1 network. Changes in receptor occupancy, for example caused by noise-like fluctuations of the extracellular cytokine concentration, which are shorter than this intrinsic response time of $\approx 15\,\text{min}$ cannot be transduced efficiently to the target genes in the nucleus. The intrinsic response time is controlled jointly by DNA binding and nuclear phosphatase activity. Interestingly, the rate constants of nuclear import and export, which strongly influence the response amplitude, have very little control on the signal duration (data not shown). By contrast, the nuclear phosphatase activity is an equally important regulator of both response amplitude and duration.

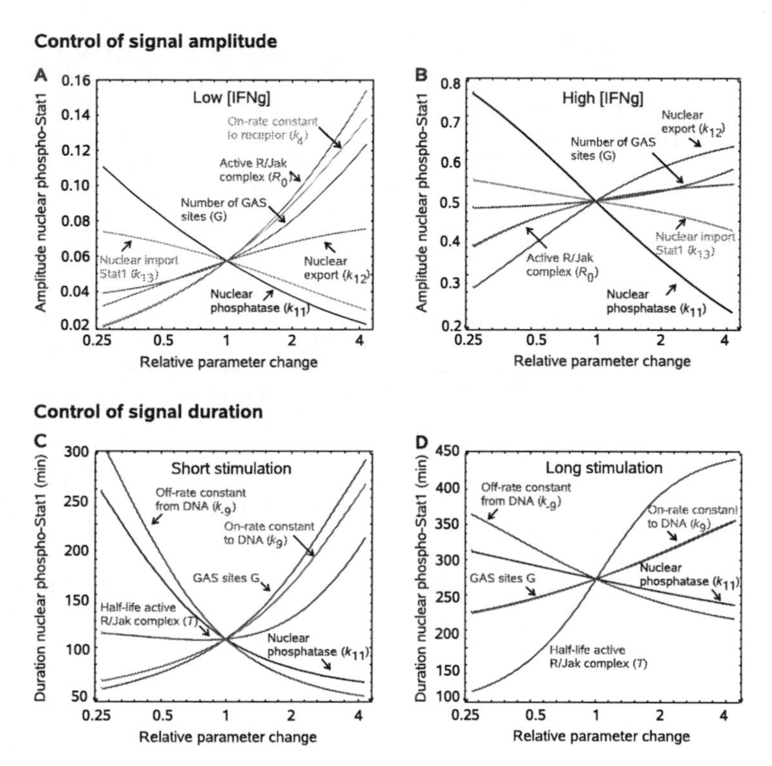

Figure 5.2: Control analysis of response amplitude and duration. Shown in parts A and B are the effect of the individual parameters on the response amplitude α by plotting the amplitude change as a function of the relative parameter variation from 0.25 fold to 4 fold of the respective reference value. Parts C and D depict the same for the response duration 2σ.

Long Stimulation

The weighting of the intrinsic and extrinsic factors is reversed for longer reference stimuli (Figure 5.2 D, $T_{0.5} = 60\,\mathrm{min}$). Now, a change in the response duration can most readily achieved by altering the stimulus time, while the system's intrinsic parameters determining DNA binding and dephosphorylation rate have a diminished control. These findings demonstrate that the IFN-γ/Stat1 pathway can function in two distinct modes. For a short stimulus, there is a one-shot phosphorylation of a certain amount of Stat1 molecules that enter the nucleus, bind to DNA and become inactivated by dephosphorylation; hence, the DNA binding and phosphatase rate parameters control the signal duration. During long-lasting stimuli, inactivated Stat1 molecules can be rephosphorylated after leaving the nucleus and thus enter repeated cycles of activation, DNA binding and inactivation. In this dynamic regime, the signal duration is largely governed by the length of the external cytokine stimulation.

5.1.3 Intrinsic Response Control

In summary, the results of our control analysis predict that intrinsic network processes of the Jak/Stat1 system play a critical role in controlling the amplitude and duration of the pathway response caused by a cytokine stimulus.

The inactivation by the nuclear phosphatase always exerts a dominant control on the amplitude and also a large control on the signal duration. It is therefore not surprising that DNA binding, which protects the phosphorylated Stat1-dimers from interaction with the phosphatase [Meyer et al., 2003, 2004], also features as a step with significant control. Moreover, the nuclear export of inactivated Stat1 strongly affects the response amplitude. Interestingly, the three controlling processes are all located after the network steps that initiate the activation cycle (phosphorylation and nuclear import).

5.2 Concentration Control Coefficients

The response of the concentrations of the various Stat pools to perturbations of individual reaction and transport rates and to stimulus variations can be analysed in a quantitative way using concentration control coefficients [Heinrich and Rapoport, 1974, Kacser and Burns, 1973]:

$$C_j^{Y_i} = \frac{\partial \ln Y_i}{\partial \ln k_j} = \frac{k_j}{Y_i}\frac{\partial Y_i}{\partial k_j}. \tag{5.2}$$

These control coefficients give the fractional change in the concentration of the Stat pool Y_i following a relative change of the model parameter k_j.

Distribution of Regulatory Control

We computed numerically the $C_j^{Y_i}$ of the extremal amplitudes for the four major Stat pools of phosphorylated and unphosphorylated Stat1 in the cytoplasm and

Process	Y_c	Y_n	Y_c^p	Y_n^p	P_n
Stimulus strength R_0	-0.664	-0.125	0.644	0.436	0.381
Receptor binding on-rate k_4	-0.643	-0.120	0.607	0.422	0.368
Receptor binding off-rate k_{-4}	0.079	0.015	-0.076	-0.052	-0.045
Phosphorylation k_5	-0.094	-0.012	0.117	0.072	0.062
Cytoplasmic phosphatase k_7	0.048	0.009	-0.036	-0.032	-0.028
Nuclear import pY-Stat k_8	0.069	0.120	-0.652	0.116	0.102
Spec. binding on-rate k_9	-0.005	-0.005	-0.001	-0.001	0.006
Spec. binding off-rate k_{-9}	0.002	0.003	0.001	0.002	-0.004
Number of GAS sites G	-0.044	-0.050	-0.001	-0.054	0.086
Unspec. binding on-Rate k_{10}	-0.042	-0.047	-0.001	-0.064	0.089
Unspec. binding off-Rate k_{-10}	0.042	0.046	0.001	0.064	-0.088
Nuclear phosphatase k_{11}	0.260	0.285	0.003	-0.619	-0.540
Nuclear export Y-Stat k_{12}	0.327	-0.677	0.402	0.357	0.312
Nuclear import Y-Stat k_{13}	-0.136	0.370	-0.342	-0.206	-0.180

Table 5.1: Concentration control coefficients of the four major Stat1 pools of cytoplasmic and nuclear Y-Stat1 and pY-Stat1 and of the maximal nuclear pY-Stat amplitude $P_n = Y_{pn} + Z_s + Z_u$. The control coefficients were computed numerically using the reference parameter set in Table 3.4 and an input stimulus function given by Equation (5.1) with $R_0 = 0.2$ and $T_{0.5} = 60\,\text{min}$.

in the nucleus (Y_c, Y_n, Y_c^p, Y_n^p) as well as the control over the total nuclear phosphorylation signal $P_n = Y_n^p + Z_s + Z_u$. The system was activated by the stimulus input function given in Equation (5.1) with an intermediate stimulation strength $R_0 = 0.2$ and a long stimulus duration $T_{0.5} = 60\,\text{min}$, the parameter values were varied by $\pm 1\%$. The corresponding values of the $C_j^{Y_i}$ are listed in Table 5.1. Please note that the normalised slopes of the amplitude response curves in Fig. 5.2 A and B are equivalent to the control coefficients $C_j^{P_n}$ of the total nuclear phospho-Stat concentration P_n.

We find for all Stat pools that the control is distributed over a multitude of different pathway steps and that some processes exert strong or medium control, while most reactions have an insignificant regulatory effect. The control coefficients of all parameters are smaller than unity, meaning that the fractional concentration change of a Stat pool after the perturbation of a network process is always smaller than the relative perturbation itself. Interestingly, the network steps with the strongest control over a specific Stat pool Y_i are all negative regulatory processes where Y_i acts as a substrate.

Process	Total Control
Stimulus strength R_0	1.869
Receptor binding on-rate k_4	1.792
Nuclear export Y-Stat k_{12}	1.763
Nuclear phosphatase k_{11}	1.167
Nuclear import Y-Stat k_{13}	1.054
Nuclear import pY-Stat k_8	0.957
Phosphorylation k_5	0.295
Receptor binding off-rate k_{-4}	0.222
Unspec. binding on-rate k_{10}	0.154
Unspec. binding off-rate k_{-10}	0.153
Number of GAS sites G	0.149
Cytoplasmic phosphatase k_7	0.125
Spec. binding on-rate k_9	0.012
Spec. binding off-rate k_{-9}	0.008

Table 5.2: Total control C_j^{Tot} of all reaction and transport processes and the stimulus strength on the four major Stat1 pools Y_c, Y_n, Y_c^p and Y_n^p.

The control coefficients of Y_n^p and P_n reflect the shape of the response curves in Figure 5.2, showing that the nuclear phosphatase has the largest negative control on the nuclear pY-Stat signal and that the receptor related parameters R_0 and k_4 and the nuclear export k_{12} are the leading regulators of nuclear phosphorylation.

Interestingly, the control coefficients of Y_n reveal that the concentration of unphosphorylated Stat in the nucleus is rather stable with respect to variations in the stimulus (as represented by the number of active receptors R_0) or to changes in the receptor on-rate constant k_4. The other three major Stat pools Y_c, Y_c^p and Y_n^p are much more sensitive to variations in these receptor related parameters and have four to five times higher control coefficients, meaning that they react stronger to changes of the stimulus strength. This homeostasis of Y_n with respect to external stimulus variations is in accordance with the numerical simulations and the model analysis regarding the relative homeostasis of nuclear Y-Stat1 presented in Section 4.2 (compare Figure 4.4).

Network Sensitivity

To quantify the sensitivity of the whole network to perturbations in a single reaction, transport step or stimulus characteristics, we computed the total network control of such a process. The total network control of a model parameter was calculated by summing the absolute values of its control coefficients over the four major Stat1 pool concentrations of phosphorylated and unphosphorylated Stat in the nucleus and in the cytoplasm:

$$C_j^{\mathrm{Tot}} = \sum_i |C_j^i| \quad \text{where } i \in \{Y_c, Y_n, Y_c^p, Y_n^p\}.$$

Table 5.2 shows a ranking of the values of the total network control C_j^{Tot} for all reaction and transport steps and for the external stimulus strength R_0. This ranking can be divided in three groups: The strongest control on the major Stat pools is exhibited by the receptor and the nuclear export of unphosphorylated Stat1, while the nuclear phosphatase and the import processes of both phosphorylated and unphosphorylated Stat1 have intermediate control. All other processes have a relatively small impact on the network. Interestingly, this ranking of the controlling processes shows the importance of the nucleo-cytoplasmic transport processes and is very similar to the control properties found for the transcriptionally active Stat1 fractions Y_n^p and P_n alone.

6

Discussion of Part I

Compartmentalisation of Activation and Inactivation Processes

The mathematical model of Jak/Stat1 signalling presented here allows to make predictions that go beyond available experimental data. Numerical simulations show that the cytoplasmic concentration of phosphorylated Stat1 remains very low after stimulation and that the cytoplasmic residence time of pY-Stat is relatively short. This residence time is largely determined by nuclear import because this process is much faster than the competing dephosphorylation reaction in the cytoplasmic compartment. Therefore there is no significant short-circuit of Stat1 activation by the cytoplasmic dephosphorylation process, and the cytoplasmic phosphatase seems to serve to avoid spontaneous phosphorylation only. Accordingly, the model simulations show that the inactivation of the system after removal of the stimulus is mainly driven by the nuclear phosphatase. Blocking rephosphorylation by a kinase inhibitor leads to a rapid collapse of the Stat1 phosphorylation signal caused by the fast nuclear dephosphorylation. We also observe in the simulations and later experimental verification measurements that the build-up and the decay of phosphorylation and nuclear accumulation proceed simultaneously.

These results show that the dynamics of the system is mainly shaped by the interplay of cytoplasmic phosphorylation, nucleo-cytoplasmic transport and nuclear dephosphorylation. Thus, the Jak/Stat network displays a clear compartmentalisation between cytoplasmic activation and nuclear deactivation. Such a network design suggests that the system response is not only determined by the interaction of kinases and phosphatases and the kinetic properties of DNA binding, but to a large extent also by the transport processes between the different cellular compartments.

Distinct Role of Shuttling of Inactive Stat1

The model simulations of the behaviour of the subcellular Stat pools during interferon induced activation reveal a high concentration and a relative homeostasis of the nuclear pool of unphosphorylated Stat1. The mathematical analysis shows that this high nuclear concentration is not caused by a slow export process - the export rate constant is nearly twice as large as the rate of nuclear dephosphorylation. The reason for the large fraction of nuclear Y-Stat1 is its constitutive import from the cytoplasm into the nucleus. In the second part of this work will show by analytical calculations how the large concentration of inactive molecules in the nucleus caused by the constitutive import process determines the high control of the export process on the pathway response we found in the control analysis.

The model also reveals that the nuclear import of Y-Stat1 serves to decouple the nuclear concentration of the phosphorylated and unphosphorylated Stat fractions by minimising the concentration changes of the inactive Stat1 pool during cytokine stimulation. Since both Stat pools may serve independent biological functions in activating gene expression, this decoupling may be biological relevant [Chatterjee-Kishore et al., 2000, Yang et al., 2005]. Taken together these results proof the important role of the shuttling process of inactive molecules for Jak/Stat1 signalling. Since dynamic shuttling is emerging to be a general property of transcription factors [Xu and Massague, 2004], these results of the theoretical analysis may also be wider applicable.

Network Control Properties Reflect Cyclical Pathway Design

The control analysis conducted for different stimulation modes reveals that the Jak/Stat signalling network does not have a hardwired stimulus-response characteristics adapted to a single function, like monitoring signal duration or achieving high signal amplification. Rather the pathway is a highly plastic system where stimulus sensitivity, signalling kinetics as well as transcriptional output are regulated by the expression levels and activities of several network components.

We find that major controlling steps of the network are located also after the initiating steps of the signalling pathway, which are phosphorylation, dimerisation and the nuclear import of the active protein. The most conspicuous result of this analysis is that the transcriptional activity of Stat1 can be strongly modulated by stimulation-independent network processes without any direct impact on the DNA binding properties of Stat1 or its interaction with other constituents of the transcriptional machinery - namely by the nuclear dephosphorylation and the import and export of the inactive molecule. The succession of tyrosine dephosphorylation and nuclear export turns out to be a critical condition for the regulation of the nuclear response, not inhibitory but stimulatory. Interestingly, the important regulatory potential of the nuclear phosphatase and of nuclear export of unphosphorylated Stat1 were both recently proofed experimentally [Lodige et al., 2005, Wang et al., 2006].

Our results on the control distribution in the network reflect the cyclical structure of the pathway: strong control of activation and deactivation, positive control of export of Stat and import of pY-Stat. Interestingly some of the controlling parameters change for weak and strong stimulation patterns.

The analysis of the regulation of the temporal characteristics of the system response revealed that the duration of the nuclear phosphorylation signal is not only dependent on the length of the extracellular stimulus but is also controlled by internal properties of the signalling network, like the characteristics of DNA binding or the nuclear dephosphorylation process. Our analysis showed that DNA binding serves as a 'retention factor' because the binding protects the phosphorylated Stat molecules from interaction with the strong phosphatase. The important role of this interplay of nuclear phosphatase and DNA binding in determining the nuclear accumulation of Stat molecules has been recently established experimentally [Meyer et al., 2003, 2004].

An interesting result of this study is that the control of the pathway steps is smaller than unity: The relative effect on the response amplitude is always smaller than the relative change in the parameter. Since this result is also valid for the receptor related processes, the pathway can not further amplify the input from the cytokine receptor by downstream signalling steps. In the following part of this work we will show by analytical calculations that this result is not caused by the specific values of the model parameters, but is a general result of the near-linearity of the system.

Our theoretical analysis predicts a large potential for the intrinsic parameters of the pathway to shape the IFN response. The model makes clear statements on how the signalling properties of the Jak/Stat network can be modulated by interfering with the kinetics of the individual steps. The predictions regarding the control capacity of the transport processes will be further addressed in part III of this thesis by experimental measurements of Stat1 transport mutants and the according theoretical analysis. Since the signalling pathways of other Stat family members show similar network structure and kinetic behaviour, our model can be used to eludicate kinetic regulation of these networks as well.

Regulatory Targets and Pathway Robustness

The results of Section 5.2 have implications for both the regulatory design and the robustness of the signal transduction pathway. For most parameters shown in Tables 5.1 and 5.2 the network can tolerate random deviations which may come about by variability in the activities or the concentrations of enzymes or of proteins of the transport machinery, caused, for example, by variability in protein expression and degradation. At the same time, it would be inefficient to target such processes by regulatory mechanisms. In contrast, the rates of Stat1 binding to the IFN-γ receptor, the amount of phosphorylated and active receptor/Jak complexes, dephosphorylation in the nucleus and the export step of the nucleo-cytoplasmic shuttling process of unphosphorylated Stat1 can be used to efficiently control the response properties of the network like the concentration of the active transcription factor in the nucleus.

Part II

Jak/Stat Core Model

7 Introduction

In part I of this work we established a detailed computational model of the Jak/Stat1 signal transduction pathway which allows us to derive experimentally testable predictions how the dynamics of the network can be modulated. To obtain a deeper understanding why certain processes are more important than others in controlling Stat1 activity, we consider a simplified variant of the Stat1 pathway model. Using the occurrence of different timescales, we reduce the complexity of the detailed Jak/Stat1 model to deduce a core model of a cyclic signalling network, which retains the fundamental network structure of the Stat1 pathway. This core model enables us to study the dynamics and behaviour of cyclic signal transduction pathways as well as the general design of such networks in a more abstract way. Furthermore, the simplified model allows for explicit analytic calculations and makes it possible to identify general properties of the system independently from particular values of the model parameters.

Based on the results of the reduction of the Jak/Stat model we then take a more abstract approach to model biological signal transduction systems. These systems normally involve the switching of signalling molecules between biological active and inactive states, for example by chemical modifications, formation of protein complexes or transport between different subcellular compartments. We model these state changes of the signalling molecules between different activity states and subcellular localisations by a network of state transitions. Using this abstract network description, we derive general rules that define which processes have a high control on the system response and where the network is robust to perturbations. That allow us to make predictions about the control distribution in signalling pathways and how specific processes influence the control of other network reactions. Applying these concepts to the Jak/Stat pathway we show how the shuttling of the inactive molecules leads to homeostasis of the nuclear pool of unphosphorylated Stat and increases the stimulus sensitivity of the active transcription factor as well as the integrated response of the network.

8 Stat1 Model Reduction

The differential equation system (3.1)–(3.10) of the detailed Jak/Stat1 pathway model already has a relatively simple mathematical structure with the majority of rate equations being linear. Exploiting the occurrence of different time scales in some network processes, the model can be reduced systematically to an even simpler and fully linear form.

8.1 Cytoplasmic Processes

The experimentally based estimate of the cytoplasmic dephosphorylation rate constant $k_7 = 0.022\,\mathrm{min}^{-1}$ shows that the dephosphorylation reaction in the cytoplasm is much slower than the nuclear import with the rate constant $k_8 = 0.3\,\mathrm{min}^{-1}$. Therefore, cytoplasmic phospho-Stat1 dimers are predominantly imported into the nucleus rather than dephosphorylated. The majority of dephosphorylation then takes place in the nucleus. Therefore the first step in the reduction of the detailed Jak/Stat1 model is to neglect the cytoplasmic phosphatase by setting $k_7 = 0$.

Next, we assume that the complex between the phosphorylated receptors and unphosphorylated Stat1 proteins formed via phospho-tyrosine SH2-domain interaction rapidly attains a steady state. This will be the case when the number of receptors and Jak kinases is sufficiently smaller than the number of Stat1 molecules. For ease of notation, we will denote the total concentration of phosphorylated receptor (including the Stat-bound fraction) by $R_T = R_p + Y_r$. Equation (3.3) together with the rate equation from Table 3.1 then implies that the enzyme-substrate complex obeys

$$Y_r = R_T \frac{Y_c}{K_M + Y_c} \tag{8.1}$$

with the half-saturation constant

$$K_M = \frac{k_{-4} + k_5}{k_4},$$

which can be interpreted as the Michaelis constant of the phosphorylation reaction. The parameter estimation shows that $K_M \gg Y_c$, so that we can approximate

$$Y_r = R_T \frac{Y_c}{K_M}.$$

It is appropriate to introduce the total cytoplasmic fraction of unphosphorylated Stat1 as a new variable:

$$S_1 = Y_c + Y_r = Y_c \left(1 + \frac{R_T}{K_M}\right). \tag{8.2}$$

Summing equations (3.3) and (3.4) and using equation (8.2), one obtains the kinetic equation

$$\dot{S}_1 = k_{12} Y_n - (k'_{13} + k'_5) S_1. \tag{8.3}$$

The new first-order rate constants k'_{13} and k'_5 are the reduced rates for nuclear import and for receptor binding of the unphosphorylated cytoplasmic Stat1 fraction:

$$k'_{13} = k_{13} \frac{K_M}{K_M + R_T}, \quad k'_5 = k_5 \frac{R_T}{K_M + R_T}. \tag{8.4}$$

8.2 Nuclear Processes

Additionally, it is assumed that both the specific (Z_s) and unspecific DNA binding (Z_u) of phosphorylated Stat1 equilibrates rapidly compared to nuclear dephosphorylation. Then we have the equilibrium relations

$$Z_s = G \frac{Y_n^p}{K_s + Y_n^p}, \quad Z_u = \frac{Y_n^p}{K_u}, \tag{8.5}$$

where

$$K_s = \frac{k_{-9}}{k_9} \quad \text{and} \quad K_u = \frac{k_{-10}}{k_{10}}$$

are the respective half-saturation constants. When unsaturated binding ($Y_n^p \ll K_s$) is assumed, the first relation simplifies to

$$Z_s = G \frac{Y_n^p}{K_s}. \tag{8.6}$$

Defining as a second new variable the total nuclear fraction of phosphorylated Stat1 dimers, in units of monomers, yields

$$S_3 = 2\,(Y_n^p + Z_s + Z_u) = 2Y_n^p(1 + \frac{G}{K_s} + \frac{1}{K_u}).$$

Summing equations (3.7), (3.9) and (3.10), one obtains the differential equation

$$\dot{S}_3 \quad = \quad 2\rho k_8 Y_c - k'_{11} S_3, \tag{8.7}$$

where ρ denotes the volume ratio of the cytoplasmic and the nuclear compartments as defined by Equation (3.11). The new kinetic parameter

$$k'_{11} \quad = \quad \frac{k_{11}}{1 + \frac{G}{K_s} + \frac{1}{K_u}} \tag{8.8}$$

is an effective dephosphorylation rate that takes into account that DNA binding protects Stat1 from dephosphorylation.

After these approximations the four variables of the reduced model are the following major Stat1 fractions:

S_1	–	cytoplasmic unphosphorylated (including receptor-bound),
Y_c^p	–	cytoplasmic phosphorylated,
S_3	–	nuclear phosphorylated (including DNA-bound) and
Y_n	–	nuclear unphosphorylated.

For reasons of symmetry we rename $2Y_c^p$ and Y_n to S_2 and S_4, respectively. Using these definitions, all Stat1 fractions are measured in monomer units, so that the conservation relation now reads

$$S_1 + S_2 + \frac{1}{\rho}\,(S_3 + S_4) = 1. \tag{8.9}$$

For ease of notation, the first-order rate constants are renamed as follows:

$\kappa_{12} = k'_5$		effective phosphorylation
$\kappa_{23} = k_8$		nuclear import of pY-Stat1
$\kappa_{34} = k'_{11}$		effective dephosphorylation
$\kappa_{41} = k_{12}$		nuclear export
$\kappa_{14} = k'_{13}$		effective nuclear import of Y-Stat1.

The equations governing the kinetics of the Stat1 signalling core model now take the form

$$\dot{S}_1 \quad = \quad \kappa_{41} S_4 - (\kappa_{12} + \kappa_{14})\, S_1 \tag{8.10}$$

$$\dot{S}_2 \quad = \quad \kappa_{12} S_1 - \kappa_{23} S_2 \tag{8.11}$$

$$\dot{S}_3 \quad = \quad \rho\kappa_{23} S_2 - \kappa_{34} S_3 \tag{8.12}$$

$$\dot{S}_4 \quad = \quad \kappa_{34} S_3 + \rho\,(\kappa_{14} S_1 - \kappa_{41} S_4)\,. \tag{8.13}$$

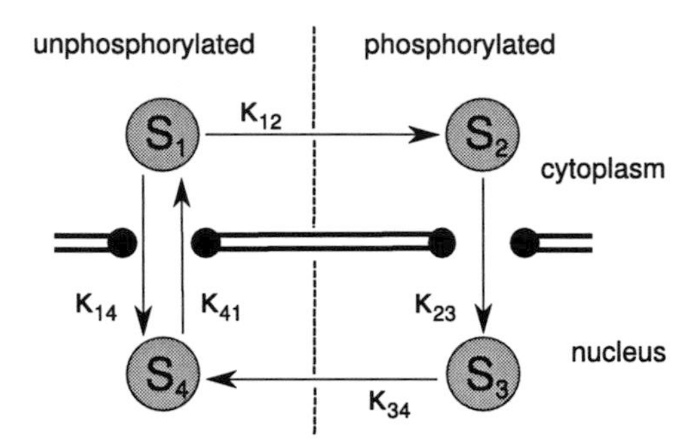

Figure 8.1: Reaction scheme of the Stat signalling core model. The reaction and transport steps are: κ_{12}: phosphorylation by Jak kinases, κ_{23}: nuclear import of phosphorylated Stat, κ_{34}: nuclear dephosphorylation, κ_{41}: nuclear export of unphosphorylated Stat, κ_{14}: nuclear import of unphosphorylated Stat.

A scheme of the core model is shown in Figure (8.1). Taken together, these simplifications have resulted in a system with linear first-order kinetics and mass conservation.

8.3 Core Model Reference Parameters

According to the results from the last section, the reference parameters of the Stat core model were calculated from the detailed model parameters given in Table 3.4 to have the following values:

$$\begin{aligned}\rho &= 1\\ \kappa_{12} &= 0.17\,\text{min}^{-1}\\ \kappa_{23} &= 0.30\,\text{min}^{-1}\\ \kappa_{41} &= 0.17\,\text{min}^{-1}\\ \kappa_{14} &= 0.12\,\text{min}^{-1}\end{aligned}$$

Comparison of numerical simulations obtained from the detailed Stat1 model and the reduced core model has shown that the effective cytoplasmic dephosphorylation according to equation (8.8) is underestimated in the simplified model, because the above assumption $Y_n \ll K_s$ is not well fulfilled in the detailed model of the system, where $K_s = 0.012$ and $Y_n^p \approx 0.1$. Therefore the numerical value for κ_{34} according to equation (8.8) was tripled to match the phosphorylation time course of the full

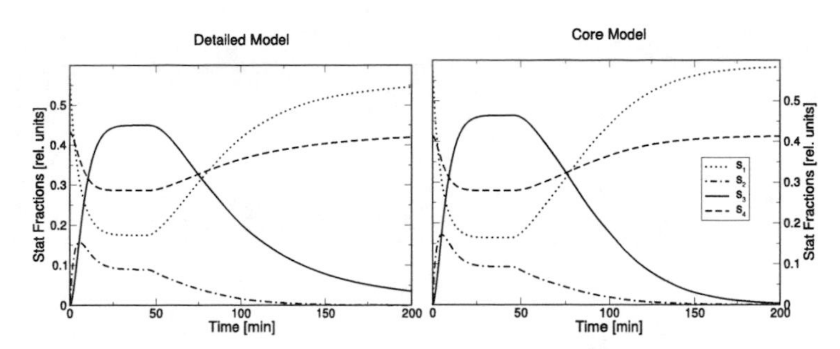

Figure 8.2: Model simulation of the activation and deactivation dynamics of the four major Stat pools in the detailed (right plot) and in the core model (left plot). Deactivation of the pathway was simulated in both models by exponentially reducing the phosphorylation rate after $t = 45$ minutes. Shown are the unphosphorylated cytoplasmic fraction (S_1, dotted line), the phosphorylated cytoplasmic Stat pool (S_2, dash-dotted line), the phosphorylated nuclear Stat pool (S_3, solid line) and the unphosphorylated nuclear fraction (S_4, dashed line).

---

model more closely, yielding $\kappa_{34} = 0.06\,\mathrm{min}^{-1}$. Additional numerical simulations have shown that this discrepancy affects only the concentration distribution among the different Stat pools S_i but not the qualitative system dynamics, the general results obtained from the linearised core model are still valid and applicable to the detailed Stat1 system. The comparison of simulated timeseries of the four major Stat pools obtained from the detailed model (3.1)–(3.10) and the simplified core model (8.10)–(8.13) shows that the dynamics of both models are quite similar (compare Figure 8.2). This reveals that the applied mathematical simplifications and assumptions are indeed valid and that the core model can serve as a good approximation for the dynamics of the detailed nonlinear system.

9

Core Model Steady State

The input stimuli of Jak/Stat pathways are typically transient. However, these stimuli may be sufficiently long that the pathway will practically reach a steady state. For example, cytokine signals between cells of the immune system are limited to certain phases of the immune responses, lasting between one and several hours. For the experimentally estimated parameters, the steady state is reached within less than 40 minutes while the deactivation happens on a much longer timescale (compare the experimental data in Figure 4.1), so that the steady state solution provides an appropriate measure of the response amplitude.

The steady state solution of the equation system (8.9)–(8.13) can be calculated using the King-Altman method (compare 10.1) or other simple algebraic methods. The steady state concentrations of the four major Stat pools of phosphorylated and unphosphorylated molecules in the nucleus or in the cytoplasm are given by

$$\overline{S}_1 = \frac{\frac{1}{\kappa_{12}}}{\frac{1}{\kappa_{12}} + \frac{1}{\kappa_{23}} + \frac{1}{\kappa_{34}} + \frac{1}{\rho\kappa_{41}}\left(1 + \frac{\kappa_{14}}{\kappa_{12}}\right)} \tag{9.1}$$

$$\overline{S}_2 = \frac{\frac{1}{\kappa_{23}}}{\frac{1}{\kappa_{12}} + \frac{1}{\kappa_{23}} + \frac{1}{\kappa_{34}} + \frac{1}{\rho\kappa_{41}}\left(1 + \frac{\kappa_{14}}{\kappa_{12}}\right)} \tag{9.2}$$

$$\overline{S}_3 = \rho \frac{\frac{1}{\kappa_{34}}}{\frac{1}{\kappa_{12}} + \frac{1}{\kappa_{23}} + \frac{1}{\kappa_{34}} + \frac{1}{\rho\kappa_{41}}\left(1 + \frac{\kappa_{14}}{\kappa_{12}}\right)} \tag{9.3}$$

$$\overline{S}_4 = \rho \frac{\frac{1}{\rho\kappa_{41}}\left(1 + \frac{\kappa_{14}}{\kappa_{12}}\right)}{\frac{1}{\kappa_{12}} + \frac{1}{\kappa_{23}} + \frac{1}{\kappa_{34}} + \frac{1}{\rho\kappa_{41}}\left(1 + \frac{\kappa_{14}}{\kappa_{12}}\right)} \tag{9.4}$$

Stat Lifetimes Determine Steady State

The concentration S_3 of the transcriptionally active nuclear phospho-Stat is the primary output signal of the pathway and determines the expression of the cytokine-induced target genes.

The steady state equation of S_3 (9.3) shown above has an intuitively appealing interpretation. We consider the average time a Stat1 molecule needs to complete one full cycle between two phosphorylation events at the receptor. This will be the sum of the average transition times between two consecutive states S_i and S_{i+1}. For example, the average time T_{23} it takes for a Stat molecule to proceed from S_2 to S_3 can be calculated as follows: Starting with a certain amount of molecules in the state S_2 and with the influx being blocked, the molecules will disappear from S_2 over time. Then the transition time T_{23} equals the lifetime τ_2 of the molecules in state S_2. The concentration decays according to $\dot{S}_2 = -\kappa_{23}S_2$. With the initial condition $S_2(0) = 1$ the transition time can be calculated formally, yielding the well-known result

$$T_{23} = \tau_2 = \frac{1}{S_2(0)} \int_0^\infty S_2(t)dt = \frac{1}{\kappa_{23}}. \tag{9.5}$$

Analogously, one finds for the transition time from S_3 to S_4

$$T_{34} = \tau_3 = \frac{1}{S_3(0)} \int_0^\infty S_3(t)dt = \frac{1}{\kappa_{34}},$$

which is identical to the lifetime τ_3 of nuclear phospho-Stat1.

Due to the bidirectional shuttling, the transition time from S_4 to S_1 is not well-defined, to close the cycle, we thus calculate how long it takes for a Stat molecule to proceed from state S_4 back to S_2. For this, equations (8.10), (8.11) and (8.13) are considered together, setting $\kappa_{34} \equiv 0$ and $\kappa_{23} \equiv 0$. Starting with the initial condition $S_4(0) = 1$, $S_1(0) = 0$ and $S_2(0) = 0$, S_2 reaches the steady state concentration $\bar{S}_2 = 1/\rho$ with the characteristic time

$$T_{42} = \frac{1}{\bar{S}_2 - S_2(0)} \int_0^\infty (\bar{S}_2 - S_2(t))dt = \frac{1}{\kappa_{12}} + \frac{1}{\rho\kappa_{41}}\left(1 + \frac{\kappa_{14}}{\kappa_{12}}\right). \tag{9.6}$$

Calculating the lifetimes of S_1 (with $S_4(0) = 0$ and $S_1(0) = 1$) and S_4 (by setting $S_4(0) = 1$ and $S_1(0) = 0$) one obtains

$$\tau_1 = \frac{1}{S_1(0)} \int_0^\infty S_1(t)dt = \frac{1}{\kappa_{12}} \tag{9.7}$$

and

$$\tau_4 = \frac{1}{S_4(0)} \int_0^\infty S_4(t)dt = \frac{1}{\rho\kappa_{41}} \left(1 + \frac{\kappa_{14}}{\kappa_{12}}\right).$$

Thus, the transition time between the states S_4 and S_2 can be written as the sum of the lifetimes of the concentrations S_4 and S_1:

$$T_{42} = \tau_4 + \tau_1.$$

Taken together, the average cycle time for a complete sequence of phosphorylation of a Stat molecule, nuclear import, DNA binding, dephosphorylation and nucleo-cytoplasmic shuttling is

$$T_C = \sum_{i=1}^{4} \tau_i = \frac{1}{\kappa_{12}} + \frac{1}{\kappa_{23}} + \frac{1}{\kappa_{34}} + \frac{1}{\rho\kappa_{41}} \left(1 + \frac{\kappa_{14}}{\kappa_{12}}\right) \tag{9.8}$$

which equals the denominator of the steady state equations (9.1) - (9.4). Inserting the reference parameter set into equation (9.8) yields for the average cycle time in the simplified core model a value of

$$T_C = 35.9\,\text{min},$$

which is in good agreement with the numeric value of 34 minutes obtained from the detailed model simulations (compare Equation 4.2).

The numerator of the system response $\overline{S}_3$ is the life-time of the phosphorylated molecule in the nucleus, T_{34}, while the factor ρ norms the nuclear concentration relative to the cytoplasmic volume. Analogous results hold for the steady-state fractions of the other states. The U3A and HeLa cell lines regarded in part I of this work have approximately equal nuclear and cytoplasmic volumes (T. MEYER, personal communication), in this case the weighting factor ρ can be omitted and equations (9.1)–(9.4) express the general result

$$\text{Steady-state fraction of Stat1} = \frac{\text{Fraction lifetime}}{\text{Cycle time}},$$

yielding a relation between the system timescales and the steady state concentrations of the network.

10

Signalling Cycles as State Transition Networks

10.1 Network of State Transitions

Signal transduction proceeds through the reversible activation of proteins by chemical modification such as phosphorylation. While one typically distinguishes between an inactive and an active state of a given signalling protein, the individual reaction steps are far more complex and may involve multiple phosphorylations, other covalent modifications, formation of protein complexes, and transport between different cellular compartments. From a more general viewpoint, such signalling pathways can be regarded as a network of state transitions, where the signalling proteins switch between various biological active or inactive forms. The Stat signalling core model (8.10) - (8.13) is an example of such a pathway, where the state network is defined by the phosphorylation state and the nuclear or cytoplasmic localisation of the Stat protein and where the transition steps are phosphorylation, dephosphorylation and transport in and out of the nucleus. In the following we will examine a generalised form of linear state transition networks and derive several properties of such networks. These general results will be applied to the Jak/Stat1 system in Chapter 11.

We consider a signalling protein that can occur in N different states. Describing the state transitions by linear rate laws, we obtain for the fractions of the various states X_i the following general system of differential equations:

$$\begin{aligned} \dot{X}_i &= k_{1i}X_1 + \ldots + k_{i-1,i}X_{i-1} - \sum_{j\neq i} k_{ij}X_i + k_{i+1,i}X_{i+1} + \ldots + k_{Ni}X_N \\ &= \sum_{j\neq i} k_{ji}X_j - \sum_{j\neq i} k_{ij}X_i \end{aligned} \qquad (10.1)$$

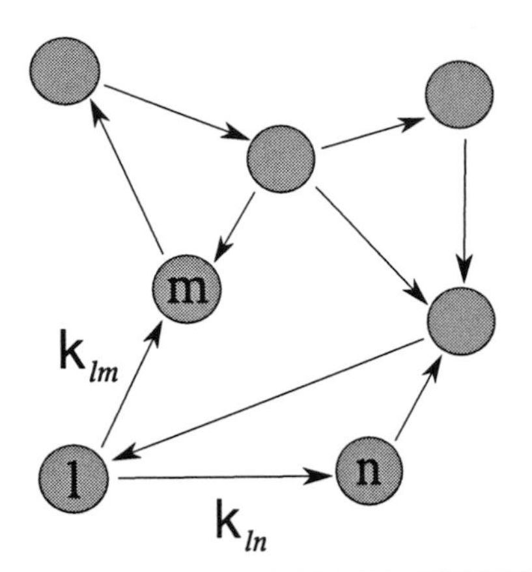

Figure 10.1: Graphical representation of a network of linear state transitions. The network nodes represent the protein concentration in a specific state while the edges between two adjacent nodes l and m stand for the the transition reaction between these two states with the transition rate k_{lm}.

where $i = 1..N$. Without loss of generality the mass conservation can be normalised to unity:

$$\sum_i X_i = 1, \tag{10.2}$$

so that the X_i are fractions of the total number of molecules.

The differential equation system (10.1) can be viewed as a graph where a node i stands for the protein concentration in state X_i and a edge k_{ij} between the two nodes i and j represents a linear transition reaction from state X_i to state X_j with the first order rate constant k_{ij} (compare Figure 10.1). Please note that self-replication reactions k_{ii} are not allowed. The same form of kinetic equations has been used for first-order enzyme reactions [King and Altman, 1956, Cornish-Bowden, 1995], early models of gene regulatory networks [Chen et al., 1999] and is a common description for ion channel gating [Koch, 1998].

Network Steady State

The steady state vector $\overline{\mathbf{X}}$ of the network defined by (10.1) is characterised by the equation system

$$\sum_{j \neq i} k_{ji}\overline{X}_j - \sum_{j \neq i} k_{ij}\overline{X}_i = 0. \tag{10.3}$$

A linear system like (10.3) can be solved analytically by ordinary algebraic methods. For linear systems which additionally exhibit a conservation relation like (10.2) King and Altman have found an algorithmic method based on the determinant method known as 'Cramer's rule' [King and Altman, 1956]. In the following paragraphs some general properties of linear state-transition networks will be derived, where we employ several arguments that have originally been developed by KING and ALTMAN in this context [King and Altman, 1956, Cornish-Bowden, 1995]. The outline of the next paragraph closely follows the derivation of the King-Altman method in Cornish-Bowden [1995], page 74, ff.

When solving the above equation system (10.3) for the steady state concentration $\overline{X}_m$ it is convenient to replace the steady state equation for X_m with the conservation relation (10.2):

$$
\begin{aligned}
-\sum k_{1j}\overline{X}_1 + k_{21}\overline{X}_2 + \ldots k_{m1}\overline{X}_m + \ldots + k_{N1}\overline{X}_N &= 0 \\
k_{12}\overline{X}_1 - \sum k_{2j}\overline{X}_2 + \ldots k_{m2}\overline{X}_m + \ldots + k_{N2}\overline{X}_N &= 0 \\
&\vdots \\
\overline{X}_1 + \overline{X}_2 + \ldots + \overline{X}_m + \ldots + \overline{X}_N &= 1 \\
&\vdots \\
k_{1N}\overline{X}_1 + k_{2N}\overline{X}_2 + \ldots k_{mN}\overline{X}_m + \ldots + -\sum k_{nj}\overline{X}_N &= 0
\end{aligned}
\tag{10.4}
$$

Cramer's Rule applied to Equations (10.4) yields

$$
\overline{X}_m = \frac{N_m}{D} = \frac{|M^m|}{|M|} \tag{10.5}
$$

where

$$
N_m := |M^m| = \begin{vmatrix}
-\sum k_{1j} & k_{21} & \ldots & 0 & \ldots & k_{N1} \\
k_{12} & -\sum k_{2j} & \ldots & 0 & \ldots & k_{N2} \\
\vdots & \vdots & & \vdots & & \vdots \\
1 & 1 & \ldots & 1 & \ldots & 1 \\
\vdots & \vdots & & \vdots & & \vdots \\
k_{1N} & k_{2N} & \ldots & 0 & \ldots & -\sum k_{Nj}
\end{vmatrix}
$$

and

$$
D := |M| = \begin{vmatrix}
-\sum k_{1j} & k_{21} & \ldots & k_{m1} & \ldots & k_{N1} \\
k_{12} & -\sum k_{2j} & \ldots & k_{m2} & \ldots & k_{N2} \\
\vdots & \vdots & & \vdots & & \vdots \\
1 & 1 & \ldots & 1 & \ldots & 1 \\
\vdots & \vdots & & \vdots & & \vdots \\
k_{1N} & k_{2N} & \ldots & k_{mN} & \ldots & -\sum k_{Nj}
\end{vmatrix}
$$

are the full and modified Cramer matrices of the linear system in Equation (10.4).

By application of $(m-1)$ switches of adjacent rows the m-th row of M, which consists entirely of ones, can be brought in the first row. Likewise, $(m-1)$ switches of adjacent columns in M^m bring the m-th column containing the zeros in the first position. Therefore it follows for the numerator N_m of (10.5)

$$\begin{aligned} N_m &= |M^m| \\ &= -1^{2(m-1)} \begin{vmatrix} 1 & 1 & 1 & \dots & 1 \\ 0 & -\sum k_{1j} & k_{21} & \dots & k_{N1} \\ 0 & k_{12} & -\sum k_{2j} & \dots & k_{N2} \\ \vdots & \vdots & \vdots & & \vdots \\ 0 & k_{1N} & k_{2N} & \dots & -\sum k_{Nj} \end{vmatrix} . \end{aligned}$$

Since the first column in the matrix shown above consists solely of zeros apart from the 1 in the first row, this 1 can be taken out as a factor, leaving a determinant of order $N-1$:

$$N_m = |M^m| = \begin{vmatrix} -\sum k_{1j} & k_{21} & \dots & k_{N1} \\ k_{12} & -\sum k_{2j} & \dots & k_{N2} \\ \vdots & \vdots & & \vdots \\ k_{1N} & k_{2N} & \dots & -\sum k_{Nj} \end{vmatrix} . \qquad (10.6)$$

Inspection of Equations (10.5) and (10.6) shows that the numerator N_m has the following properties (for details see Cornish-Bowden [1995]):

(I) Because $\sum_i X_i = 1$ it follows immediately that

$$\sum_i N_i = D. \qquad (10.7)$$

(II) It can be shown that the expansion of the determinant $|M^m|$ is a sum of products with the same sign, and this sign is the same for all m. The determinant is positive if its order $(N-1)$ is even and negative otherwise. For odd $(N-1)$ each N_m can be multiplied by -1 without changing the value of $\frac{N_m}{D}$; therefore N_m can always be chosen to be positive.

(III) Each term in the expansion of $|M^m|$ is a product which contains exactly one matrix element M^m_{ij} from each column j. After some algebra, one finds, moreover, that the sum obtained from the expansion of $|M^m|$ contains each k_{ij} for each i exactly once. Thus, N_m can always be expressed as

$$N_m = \sum_j k_{ij} \frac{\partial N_m}{\partial k_{ij}} \quad \text{for all } i \neq m. \qquad (10.8)$$

(IV) $|M^m|$ contains no constants k_{mi} with m as the first index. Hence its expansion N_m also does not contain the reaction rate k_{mi} and therefore

$$\frac{\partial N_m}{\partial k_{mi}} = 0 \quad \text{for all } i. \tag{10.9}$$

The King-Altman Method

KING and ALTMAN have found a graph-theoretical method to compute the steady state (10.5) in an algorithmic way from the topology of the network without explicitly evaluating the determinants $|M^m|$ and $|M|$ [King and Altman, 1956]. To apply the King-Altman method for deriving the expression of N_m it is necessary to find every allowed sub-path in the network which satisfies the following rules:

1. An allowed sub-path contains $N-1$ edges,
2. obeys the direction of these edges,
3. visits every network node once (i.e. contains no closed loops) and
4. ends exclusively at node m.

Each allowed sub-path is then interpreted as a product $\mathcal{S}_i$ of the rate constants k_{ij} corresponding to the edges in the sub-path. The numerator $|M^m| = N_m$ of Equation (10.5) is the sum of all products $\mathcal{S}_i$ corresponding to all allowed sub-paths to node m. The denominator $|M| = D$ can then easily calculated from (10.7) as the sum of all allowed subpaths of the network:

$$D = \mathcal{S}^T = \sum \mathcal{S}_i \,.$$

The rules (I) – (IV) and the King-Altman algorithm imply the following four inequalities:

- With property (II) and the fact that all $k_{ij} \geq 0$ follows

$$\frac{\partial N_m}{\partial k_{ij}} \geq 0 \quad \text{for all } i,j,m. \tag{10.10}$$

- From (III) and Equation (10.10) we have

$$\frac{\partial N_m}{\partial k_{ij}} k_{ij} \leq N_m \quad \text{for all } i,j,m, \tag{10.11}$$

  implying that

$$\frac{\partial D}{\partial k_{ij}} k_{ij} \overset{(10.7)}{=} \sum_l \frac{\partial N_l}{\partial k_{ij}} k_{ij} \leq D \quad \text{for all } i,j. \tag{10.12}$$

- From the King-Altman algorithm follows that the expression for D must contain each k_{ij} at least once; since D is a sum of positive products this implies that
$$\frac{\partial D}{\partial k_{ij}} > 0 \quad \text{for all } i,j. \tag{10.13}$$

In the calculations of the next sections these properties will be used.

10.2 Concentration and Flux Control Coefficients

Node Concentration Control

The concept of concentration control coefficients has already been introduced in Section 5.2. In the case presented here, the coefficient C_{ij}^m measures how a change in the rate parameter k_{ij} affects the steady state concentration $\overline{X}_m$. C_{ij}^m can be calculated as follows

$$\begin{aligned} C_{ij}^m &:= \frac{k_{ij}}{\overline{X}_m}\frac{\partial \overline{X}_m}{\partial k_{ij}} \\ &= \frac{k_{ij}}{N_m D^{-1}}\frac{\partial (N_m D^{-1})}{\partial k_{ij}} \\ &= \frac{k_{ij}}{N_m}\frac{\partial N_m}{\partial k_{ij}} - \frac{k_{ij}}{D}\frac{\partial D}{\partial k_{ij}}. \end{aligned}$$

We define the *node concentration control* $\widehat{C}_i^m$ by summing C_{ij}^m over all edges which originate at node i, which yields

$$\widehat{C}_i^m = \sum_j C_{ij}^m = \sum_j \frac{k_{ij}}{N_m}\frac{\partial N_m}{\partial k_{ij}} - \sum_j \frac{k_{ij}}{D}\frac{\partial D}{\partial k_{ij}}.$$

The node concentration control $\widehat{C}_i^m$ measures the total control of all processes emanating from node i over the concentration of state m (compare Figure 10.2).

For the first sum in the above definition of $\widehat{C}_i^m$ we can write

$$\begin{aligned} \sum_j \frac{k_{ij}}{N_m}\frac{\partial N_m}{\partial k_{ij}} &= \begin{cases} \frac{1}{N_m}\sum_j k_{mj}\frac{\partial N_m}{\partial k_{mj}} = 0 & \text{for } i = m, \text{ cf. (10.9)} \\ \frac{1}{N_m}\sum_j k_{ij}\frac{\partial N_m}{\partial k_{ij}} = 1 & \text{for } i \neq m, \text{ cf. (10.8)} \end{cases} \\ &= 1 - \delta_{im}, \end{aligned}$$

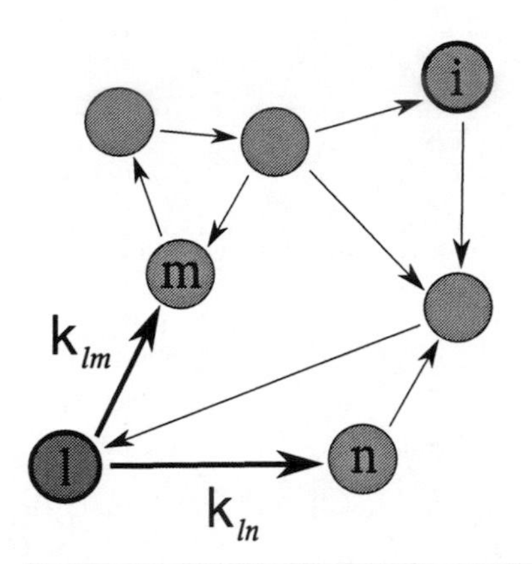

Figure 10.2: Illustration of the concept of node concentration control $\widehat{C}_l^i$. $\widehat{C}_l^i$ denotes the total control of all reactions emanating from node l (red) over the steady state concentration of the molecules in state i (blue). In the shown example this would be the sum of the two control coefficients $C_{lm}^i + C_{ln}^i$.

where δ_{im} denotes the Kronecker symbol. Using Equations (10.9) and (10.8) again, the second sum can be expressed as

$$\begin{aligned}
\sum_j \frac{k_{ij}}{D}\frac{\partial D}{\partial k_{ij}} &\overset{(10.7)}{=} \frac{1}{D}\sum_j k_{ij}\frac{\partial(\sum_l N_l)}{\partial k_{ij}} \\
&= \frac{1}{D}\sum_l\sum_j k_{ij}\frac{\partial N_l}{\partial k_{ij}} \\
&\overset{(10.9),(10.8)}{=} \frac{1}{D}\sum_{l\neq i} N_l \\
&= \frac{D-N_i}{D} \\
&= 1 - X_i\,.
\end{aligned}$$

Taken together, this yields for the total control of all reactions emanating from node i on the concentration $\overline{X}_m$

$$\begin{aligned}
\widehat{C}_i^m = \sum_j C_{ij}^m &= \begin{cases} X_m - 1 & \text{for } i = m \\ X_i & \text{for } i \neq m \end{cases} \\
&= X_i - \delta_{im}\,. \qquad (10.14)
\end{aligned}$$

Please note that by summing over all nodes i one immediately obtains the summation theorem for concentration control coefficients of metabolic control theory

[Heinrich and Schuster, 1996]:

$$\sum_i \widehat{C}_i^m = \sum_i (X_i - \delta_{im}) = \left(\sum_i X_i\right) - 1 = 0.$$

In the above derivation a general network design was assumed, in which every pair of nodes i and j is connected by two edges k_{ij} and k_{ji}. More specific network topologies where not all edges are present can easily be obtained by setting the missing rate constants to zero, which will not change the presented results.

Node Flux Control

The steady state flux between two adjacent nodes m and n is given by [Heinrich and Schuster, 1996]:

$$J_{mn} = k_{mn}\overline{X}_m - k_{nm}\overline{X}_n$$

and the steady state flux control coefficient of the rate k_{ij} on the flux J_{mn} is defined as

$$C_{ij}^{mn} = \frac{k_{ij}}{J_{mn}}\frac{\partial J_{mn}}{\partial k_{ij}} = \frac{k_{ij}}{J_{mn}}\frac{\partial}{\partial k_{ij}}\left(k_{mn}\overline{X}_m - k_{nm}\overline{X}_n\right).$$

The flux control coefficient can be expressed in terms of concentration control coefficients introduced above:

$$\begin{aligned}
C_{ij}^{mn} &= \frac{k_{ij}}{J_{mn}}\frac{\partial}{\partial k_{ij}}\left(k_{mn}\overline{X}_m - k_{nm}\overline{X}_n\right) \\
&= \frac{k_{ij}}{J_{mn}}\left(\frac{\partial \overline{X}_m}{\partial k_{ij}}k_{mn} + \frac{\partial k_{mn}}{\partial k_{ij}}\overline{X}_m - \frac{\partial \overline{X}_n}{\partial k_{ij}}k_{nm} - \frac{\partial k_{nm}}{\partial k_{ij}}\overline{X}_n\right) \\
&= \frac{k_{ij}}{J_{mn}}\left(\frac{\overline{X}_m}{k_{ij}}C_{ij}^m\, k_{mn} - \frac{\overline{X}_n}{k_{ij}}C_{ij}^n\, k_{nm} + \delta_{im}\delta_{jn}\overline{X}_m - \delta_{in}\delta_{jm}\overline{X}_n\right) \\
&= \frac{\overline{X}_m}{J_{mn}}C_{ij}^m\, k_{mn} - \frac{\overline{X}_n}{J_{mn}}C_{ij}^n\, k_{nm} + \frac{k_{ij}}{J_{mn}}\left(\delta_{im}\delta_{jn}\overline{X}_m - \delta_{in}\delta_{jm}\overline{X}_n\right),
\end{aligned}$$

where we used the relation

$$\frac{\partial k_{mn}}{\partial k_{ij}} = \delta_{im}\delta_{jn}\,.$$

Similar to the previous section we define the *node flux control* $\widetilde{C}_i^{mn}$ as a measure of the total control of all processes emanating from node i on the flux between the compounds m and n. Hence, we sum C_{ij}^{mn} over all edges which originate at node

i, which yields

$$\begin{aligned}
\widetilde{C}_i^{mn} &= \sum_j C_{ij}^{mn} \\
&= \frac{\overline{X}_m}{J_{mn}} k_{mn} \sum_j C_{ij}^m - \frac{\overline{X}_n}{J_{mn}} k_{nm} \sum_j C_{ij}^n + \\
&\quad \frac{1}{J_{mn}} \sum_j k_{ij} \left(\delta_{im}\delta_{jn}\overline{X}_m - \delta_{in}\delta_{jm}\overline{X}_n\right) \\
&= \frac{\overline{X}_m}{J_{mn}} k_{mn}(\overline{X}_i - \delta_{im}) - \frac{\overline{X}_n}{J_{mn}} k_{nm}(\overline{X}_i - \delta_{in}) + \\
&\quad \frac{1}{J_{mn}} \sum_j k_{ij} \left(\delta_{im}\delta_{jn}\overline{X}_m - \delta_{in}\delta_{jm}\overline{X}_n\right),
\end{aligned}$$

where we have used the node concentration control rule derived in the previous section. One can distinguish three different cases:

1) $m, n \neq i$

$$\begin{aligned}
\widetilde{C}_i^{mn} &= \sum_j C_{ij}^{mn} \\
&= \frac{\overline{X}_m k_{mn}}{J_{mn}} \overline{X}_i - \frac{\overline{X}_n k_{nm}}{J_{mn}} \overline{X}_i \\
&= \overline{X}_i \left(\frac{\overline{X}_m k_{mn} - \overline{X}_n k_{nm}}{J_{mn}}\right) \\
&= \overline{X}_i
\end{aligned}$$

2) $m = i, n \neq i$

$$\begin{aligned}
\widetilde{C}_i^{in} &= \sum_j C_{ij}^{in} \\
&= \frac{\overline{X}_i k_{in}}{J_{in}} (\overline{X}_i - 1) - \frac{\overline{X}_n k_{ni}}{J_{in}} \overline{X}_i + \frac{1}{J_{in}} \sum_j k_{ij}\, \delta_{jn} \overline{X}_i \\
&= \frac{\overline{X}_i k_{in}}{J_{in}} (\overline{X}_i - 1) - \frac{\overline{X}_n k_{ni}}{J_{in}} \overline{X}_i + \frac{\overline{X}_i k_{in}}{J_{in}} \\
&= \overline{X}_i \left(\frac{\overline{X}_i k_{in} - \overline{X}_n k_{ni}}{J_{in}}\right) \\
&= \overline{X}_i
\end{aligned}$$

3) $m \neq i, n = i$

$$
\begin{aligned}
\widetilde{C}_i^{mi} &= \sum_j C_{ij}^{mi} \\
&= \frac{\overline{X}_m k_{mi}}{J_{mi}} \overline{X}_i - \frac{\overline{X}_i k_{im}}{J_{mi}} (\overline{X}_i - 1) - \frac{1}{J_{mi}} \sum_j k_{ij} \delta_{jm} \overline{X}_i \\
&= \frac{\overline{X}_m k_{mi}}{J_{mi}} \overline{X}_i - \frac{\overline{X}_i k_{im}}{J_{mi}} (\overline{X}_i - 1) - \frac{\overline{X}_i k_{im}}{J_{mi}} \\
&= \overline{X}_i \left(\frac{\overline{X}_m k_{mi} - \overline{X}_i k_{im}}{J_{mi}} \right) \\
&= \overline{X}_i
\end{aligned}
$$

Thus, we have

$$
\widetilde{C}_i^{mn} = \sum_j C_{ij}^{mn} = \overline{X}_i \quad \text{for all } i, m, n. \tag{10.15}
$$

This relation implies that the total control of all processes for which node i is a substrate over the flux between two arbitrary adjacent nodes is equal to the relative steady state concentration of node X_i.

Network State Occupancies Determine Control Distribution

Equations (10.14) and (10.15) reveal important relations between the steady state concentrations of the system and the flux and concentration control coefficients of the network. The total control of all reactions emanating from a node i onto all network fluxes J_{mn} and on the concentration at an arbitrary different node X_m is equal to the relative steady-state concentration at node i. Furthermore, the total control of all reaction steps emanating from a node i onto the concentration of their substrate X_i is equal to the negative sum of the steady state concentrations of the remaining compounds $-\sum_{l \neq i} \overline{X}_l$. Thus, the results above reveal a novel relation between the occupancies of the network states and the distribution of control among the network processes. Please note that Equation (10.14) also implies that the concentration control of a reaction which is the only outgoing edge from a specific node can be determined experimentally by simply measuring the relative steady-state concentration of that node.

Concentration Control is Bounded by Unity

Since N_m is a positive sum of products of the k_{ij} (see property (II) on page 68) it follows from equation (10.8) that

$$0 \leq k_{ij} \frac{\partial N_m}{\partial k_{ij}} \leq N_m \,, \tag{10.16}$$

therefore

$$0 \leq \frac{k_{ij}}{N_m} \frac{\partial N_m}{\partial k_{ij}} \leq 1 \,, \tag{10.17}$$

and with Equation (10.7)

$$0 \leq \frac{k_{ij}}{D} \frac{\partial D}{\partial k_{ij}} = \frac{1}{D} \sum_l k_{ij} \frac{\partial N_l}{\partial k_{ij}} \leq 1 \,. \tag{10.18}$$

Then the minimum of the concentration control coefficient is given by

$$\min C_{ij}^m = \min \left(\frac{k_{ij}}{N_m} \frac{\partial N_m}{\partial k_{ij}} - \frac{k_{ij}}{D} \frac{\partial D}{\partial k_{ij}} \right) = -1 \tag{10.19}$$

and the maximum control is

$$\max C_{ij}^m = \max \left(\frac{k_{ij}}{N_m} \frac{\partial N_m}{\partial k_{ij}} - \frac{k_{ij}}{D} \frac{\partial D}{\partial k_{ij}} \right) = 1 \,. \tag{10.20}$$

Therefore the concentration control of all reactions in a linear state transition network such as Equation (10.1) will never exceed unity:

$$|C_{ij}^m| \leq 1 \,, \tag{10.21}$$

so that the relative effect of a perturbation on a network process is always smaller than the relative perturbation. This result is in accordance with the numerical control analysis of the detailed Jak/Stat1 model in Section 5.2 and explains the previous finding of the subsensitivity of the Stat1 network by its near-linearity.

10.3 Network Topology Determines Control

Whether a given edge in a network of state transitions has a positive or negative control on the concentration of a specific state may depend on the values of the network rate constants k_{ij}. However, under certain conditions the sign of concentration control coefficients may be parameter independent, solely determined by the network topology. The knowledge of the sign of control of given reactions independent from a specific network parameter set turns out to be helpful for the control analysis of particular networks, for example for the following analysis of the Jak/Stat system. Therefore some topology related rules determining this sign will be derived in the following paragraphs.

Steps with Negative Control

Since k_{ij} and N_m are positive (see property II on page 68) the sign of

$$C_{ij}^m = \frac{k_{ij}}{N_m}\left(\frac{\partial N_m}{\partial k_{ij}} - \frac{N_m}{D}\frac{\partial D}{\partial k_{ij}}\right) \tag{10.22}$$

is determined by the value of

$$\frac{\partial N_m}{\partial k_{ij}} - \frac{N_m}{D}\frac{\partial D}{\partial k_{ij}}, \tag{10.23}$$

in which the last term is always positive (compare equation (10.13)).

If in a given network no allowed sub-path leading to node m contains the edge k_{ij} it follows immediately that

$$\frac{\partial N_m}{\partial k_{ij}} = 0 \tag{10.24}$$

and therefore

$$C_{ij}^m < 0 \tag{10.25}$$

in such a network.

Steps with Positive Control

If in a given network all allowed sub-paths leading to node m contain the edge k_{ij} then

$$\frac{\partial N_m}{\partial k_{ij}} k_{ij} = N_m \tag{10.26}$$

and

$$\begin{aligned} \frac{\partial N_m}{\partial k_{ij}} - \frac{N_m}{D}\frac{\partial D}{\partial k_{ij}} &= \frac{\partial N_m}{\partial k_{ij}} - \frac{\frac{\partial N_m}{\partial k_{ij}} k_{ij}}{D}\frac{\partial D}{\partial k_{ij}} \\ &= \frac{\partial N_m}{\partial k_{ij}}\left(1 - \frac{k_{ij}}{D}\frac{\partial D}{\partial k_{ij}}\right). \end{aligned} \tag{10.27}$$

Substituting Equations (10.12) and (10.10) in Equation (10.27) yields

$$C_{ij}^m > 0. \tag{10.28}$$

Furthermore, if all allowed sub-paths of the network through the edge k_{ij} lead exclusively to node m, it follows from (10.7) that

$$\frac{\partial D}{\partial k_{ij}} = \frac{\partial N_m}{\partial k_{ij}}.$$

Then, we have for

$$\begin{aligned} \frac{\partial N_m}{\partial k_{ij}} - \frac{N_m}{D}\frac{\partial D}{\partial k_{ij}} &= \frac{\partial N_m}{\partial k_{ij}} - \frac{N_m}{D}\frac{\partial N_m}{\partial k_{ij}} \\ &= \frac{\partial N_m}{\partial k_{ij}}\left(1 - \frac{N_m}{D}\right), \end{aligned} \tag{10.29}$$

where

$$\frac{N_m}{D} < 1 .$$

From (10.10), it follows again for such a reaction k_{ij}

$$C_{ij}^m > 0 . \tag{10.30}$$

Positive Exceeds Negative Control

The sign of the control of the edges leading out of a specific node on different compounds can be positive or negative. If the edges are grouped according to the sign of their control the node concentration control summation rule can be written as

$$\widehat{C}_i^m = \sum_j C_{ij}^m = \sum_{j \in J^+} C_{ij}^m - \sum_{j \in J^-} |C_{ij}^m| = \overline{X}_i \quad \text{for } i \neq m , \tag{10.31}$$

where the first sum contains all edges with positive sign while the second sum contains the negative ones.

Since the concentration $\overline{X}_i$ must be positive, it follows immediately from this notation that the total control of all positive edges must exceed the magnitude of the total control of the negative ones. For $i = m$ this result is inverted , because then $\widehat{C}_i^i = \overline{X}_i - 1$ will be negative and the magnitude of the total control of all negative edges must exceed the total control of the edges with positive control.

11

Core Model Control Analysis

Having derived several general properties of linear state transition networks regarding the distribution of control among the individual reaction steps and the magnitude and sign of their control coefficients, we can return to the Jak/Stat signalling core model and apply the results of the previous chapter.

11.1 Control of System Response

The primary output of the cyclic Stat signal transduction network is the nuclear concentration of the active transcription factor S_3. Hence, we can use the steady state equation of $\overline{S}_3$, Equation (9.3) to characterise how the system response in terms of nuclear pY-Stat depends on the parameters of the individual reaction and transport processes in the cycle. This can be done in a systematic way using the concept of concentration control coefficients already introduced in the previous chapters:

$$C_{ij}^{m} = \frac{\partial \ln \overline{S}_m}{\partial \ln \kappa_{ij}} = \frac{\kappa_{ij}}{\overline{S}_m} \frac{\partial \overline{S}_m}{\partial \kappa_{ij}}. \tag{11.1}$$

With the help of the summation rules for the node concentration control $\widehat{C}_i^m$ derived in Chapter 10 and the steady-state solution (9.1)–(9.4), the control coefficients of the processes which do not share their substrate with other reactions can be directly read off the reaction scheme in Figure 8.1. These three control coefficients for nuclear phospho-Stat1 S_3 are given by:

$$\begin{aligned} C_{23}^3 &= \overline{S_2} &&= \tfrac{1}{T_C}\tfrac{1}{\kappa_{23}} \\ C_{34}^3 &= \overline{S_3} - 1 &&= -\tfrac{1}{T_C}\left(\tfrac{1}{\kappa_{12}} + \tfrac{1}{\kappa_{23}} + \tfrac{1}{\kappa_{41}}\left(1 + \tfrac{\kappa_{14}}{\kappa_{12}}\right)\right) \\ C_{41}^3 &= \overline{S_4} &&= \tfrac{1}{T_C}\tfrac{1}{\kappa_{41}}\left(1 + \tfrac{\kappa_{14}}{\kappa_{12}}\right) \end{aligned} \tag{11.2}$$

For the kinase reaction and the nuclear import process of unphosphorylated Stat1 the summation rule yields

$$C_{12}^3 + C_{14}^3 = \overline{S_1} = \frac{1}{T_C}\frac{1}{\kappa_{12}}$$

where we have

$$\begin{aligned} C_{12}^3 &= \frac{1}{T_C}\frac{1}{\kappa_{12}}\left(1 + \frac{\kappa_{14}}{\kappa_{41}}\right) \\ C_{14}^3 &= -\frac{1}{T_C}\frac{\kappa_{14}}{\kappa_{12}\kappa_{41}}. \end{aligned}$$

The expressions for the control coefficients reveal that each coefficient retains the same sign for all values of the kinetic parameters. This implies that whether a process exerts a positive or a negative effect on the nuclear accumulation of phospho-Stat1 depends on its position and direction in the cycle, but not on the specific parameters. Phosphorylation, nuclear import of the phospho-protein and nuclear export always exert positive control (C_{12}^3, C_{23}^3 and $C_{41}^3 > 0$), while nuclear dephosphorylation and import of the unphosphorylated species have negative control (C_{34}^3 and $C_{14}^3 < 0$).

The summation rule of the node concentration control states that the magnitude of control is related to the steady-state pool size. Nuclear unphosphorylated Stat1 and cytoplasmic pY-Stat1 serve as substrates for a single reaction only, thus nuclear import of phosphorylated and export of unphosphorylated Stat1 exert control proportional to the amount of their respective substrates, $C_{23}^3 = \overline{S}_2$ and $C_{41}^3 = \overline{S}_4$. Similarly, the magnitude of the control of the nuclear dephosphorylation reaction $C_{34}^3 = \overline{S}_3 - 1$ is reduced with growing phosphatase substrate concentration $\overline{S}_3$. Thus, changes in the rate parameters leading to different steady state concentrations will also directly change the control of the dephosphorylation reaction and of the import and export reaction.

Table 11.1 summarises the analytical expressions of the control coefficients of S_3 and additionally lists the coefficients for the remaining three Stat fractions. The corresponding numerical values of the coefficients for the core model reference parameters are listed in Table 11.2.

	C_{12}	C_{23}	C_{34}	C_{41}	C_{14}
S_1	$\left(1+\frac{\kappa_{14}}{\kappa_{41}}\right)\overline{S}_1 - 1$	$\overline{S}_2$	$\overline{S}_3$	$\overline{S}_4$	$-\frac{\kappa_{14}}{\kappa_{41}}\overline{S}_1$
S_2	$\left(1+\frac{\kappa_{14}}{\kappa_{41}}\right)\overline{S}_1$	$\overline{S}_2 - 1$	$\overline{S}_3$	$\overline{S}_4$	$-\frac{\kappa_{14}}{\kappa_{41}}\overline{S}_1$
S_3	$\left(1+\frac{\kappa_{14}}{\kappa_{41}}\right)\overline{S}_1$	$\overline{S}_2$	$\overline{S}_3 - 1$	$\overline{S}_4$	$-\frac{\kappa_{14}}{\kappa_{41}}\overline{S}_1$
S_4	$\left(1+\frac{\kappa_{14}}{\kappa_{41}}\right)\overline{S}_1 - \frac{\kappa_{14}}{\kappa_{12}+\kappa_{14}}$	$\overline{S}_2$	$\overline{S}_3$	$\overline{S}_4 - 1$	$\frac{\kappa_{14}}{\kappa_{12}+\kappa_{14}} - \frac{\kappa_{14}}{\kappa_{41}}\overline{S}_1$

Table 11.1: Analytical expressions for the control coefficients of all four Stat pools of the Stat core model.

11.2 Phosphatase has Maximal Control

From equations (11.2) follows that the magnitude of the nuclear phosphatase reaction can be written as

$$\begin{aligned} \left|C^3_{34}\right| &= \overline{S_1} + \overline{S_2} + \overline{S_4} \\ &= \overline{S_1} + \left|C^3_{23}\right| + \left|C^3_{41}\right| . \end{aligned} \tag{11.3}$$

This equation proofs that

$$\left|C^3_{34}\right| > \left|C^3_{23}\right| \quad \text{and} \quad \left|C^3_{34}\right| > \left|C^3_{41}\right| . \tag{11.4}$$

According to the King-Altman algorithm introduced in Section 10.1, there is only one allowed network sub-path leading to S_3 and this path must contain the reaction κ_{12} (compare the reaction scheme in Figure 8.1). Thus the sign rule (10.28) states that $C^3_{12} > 0$. On the contrary, no allowed sub-path to S_3 can contain the edge κ_{14}, therefore it follows immediately that $|C^3_{14}| < 0$, see equation (10.25). Furthermore, Equation (10.31) proofs that the positive control of the receptor kinase exceeds the magnitude of the negative control of the import process of inactive Stat:

$$\left|C^3_{12}\right| > \left|C^3_{14}\right| . \tag{11.5}$$

Inspection of the network topology shows that the only sub-path product $\mathcal{S}_i$ which can contain the reaction κ_{14} leads to S_4, thus

$$\frac{\partial \mathcal{S}_i}{\partial \kappa_{14}} = 0 \quad \text{for } i \neq 4,$$

and

$$\frac{\partial \mathcal{S}_i}{\partial \kappa_{14}} \neq 0 \quad \text{for } i = 4.$$

Table 11.2: Numerical control coefficient values (calculated for the reference parameter values of the Stat core model as listed in section 8.3)

	S_1	S_2	S_3	S_4
C_{12}	-0.721	0.279	0.279	-0.134
C_{23}	0.093	-0.907	0.093	0.093
C_{34}	0.464	0.464	-0.536	0.464
C_{41}	0.279	0.279	0.279	-0.721
C_{14}	-0.116	-0.116	-0.116	0.298

This implies for the derivative of the sum over all allowed sub-path products $\mathcal{S}^T$:

$$\frac{\partial \mathcal{S}^T}{\partial \kappa_{14}} = \frac{\partial \sum_i \mathcal{S}_i}{\partial \kappa_{14}} = \sum_i \frac{\partial \mathcal{S}_i}{\partial \kappa_{14}} = \frac{\partial \mathcal{S}_4}{\partial \kappa_{14}}.$$

With this expression and the fact that

$$S_3 = \frac{\mathcal{S}_3}{\mathcal{S}^T}$$

an inequality for the magnitude of the import control can be written as (compare Equation (10.11))

$$\begin{aligned} \left|C_{14}^3\right| &= \left|\frac{\kappa_{14}}{S_3}\frac{\partial S_3}{\partial \kappa_{14}}\right| = \left|\frac{\kappa_{14}}{\mathcal{S}_3}\left(\frac{\partial \mathcal{S}_3}{\partial \kappa_{14}} - \frac{\mathcal{S}_3}{\mathcal{S}^T}\frac{\partial \mathcal{S}^T}{\partial \kappa_{14}}\right)\right| = \frac{\kappa_{14}}{\mathcal{S}^T}\frac{\partial \mathcal{S}_4}{\partial \kappa_{14}} \\ &\leq \frac{\mathcal{S}_4}{\mathcal{S}^T} = \left|C_{41}^3\right|. \end{aligned}$$

Since the kinase control can be expressed as

$$C_{12}^3 = \left|C_{12}^3\right| = \overline{S_1} + \left|C_{14}^3\right|,$$

the above inequality together with equation (11.3) implies that

$$\left|C_{34}^3\right| > \left|C_{12}^3\right|. \tag{11.6}$$

Equations (11.4), (11.5) and (11.6) proof that C_{34}^3 always has the largest absolute value of all control coefficients, therefore nuclear dephosphorylation is the step with the strongest control independently of the kinetic parameter values. This analytic result shows that the dominant control of nuclear dephosphorylation over the nuclear concentration of the transcriptionally active pY-Stat which has been found in the numerical control analysis of the detailed Jak/Stat1 model in Chapter 5.2 represents a fundamental structural property of the Stat signalling network.

11.3 Effects of Shuttling

The discovery of an import pathway for unphosphorylated Stat1 (rate κ_{14} in Figure 8.1) has raised the question of its functional significance [Meyer et al., 2002b]. In the cycle of Stat1 activation/inactivation, this step would seem dispensable; instead a Stat pathway design lacking this import step has originally been anticipated [Swameye et al., 2003].

Nuclear import of Y-Stat will ensure the cytokine independent nuclear presence of the unphosphorylated molecule, which may serve specific functions [Chatterjee-Kishore et al., 2000, Yang et al., 2005]. However, we have found that the continuous nucleo-cytoplasmic shuttling of unphosphorylated Stat1 that proceeds through the steps with rate constants κ_{41} and κ_{14} has critical implications for the regulation of both the unphosphorylated and phosphorylated Stat1 fraction. To demonstrate this, we will compare the wild-type pathway design (Figure 8.1) with a hypothetical pathway shown in Figure 11.1, in which nuclear import of unphosphorylated Stat1, and thus nucleo-cytoplasmic shuttling, does not occur.

The steady state equations (9.1)–(9.4) of the wild-type pathway show that a similar signal transduction system can be constructed by removing the nuclear import κ_{14} of unphosphorylated Stat S_1 and simultaneously reducing the value of the export rate constant κ_{41}. If the reduced export rate κ'_{41} is chosen as

$$\kappa'_{41} = \kappa_{41} \frac{\kappa_{12}}{\kappa_{12} + \kappa_{14}} \qquad (11.7)$$

both systems exhibit identical steady state concentrations for all four S_i. Therefore this hypothetical pathway can serve as a reference system to analyse the effect on the system dynamics exerted by the shuttling of unphosphorylated Stat molecules. For ease of notation the hypothetical system will be called ΔI throughout this section.

Homeostasis of Unphosphorylated Stat1 in the Nucleus

The summation rule for the node concentration control (10.14) allows for a general comparison of the control of specific steps in the hypothetical and wild-type Stat pathways. Since both pathways exhibit an identical steady state $\overline{S}$ the sum at one node S_i

$$\widehat{C}_i^m = \sum_j C_{ij}^m = \overline{S}_i - \delta_{i,m}$$

must yield the same value in both systems.

If the control of the receptor-associated Jak kinase κ_{12} on the concentration of unphosphorylated nuclear Stat S_4 is considered, the above summation rule for

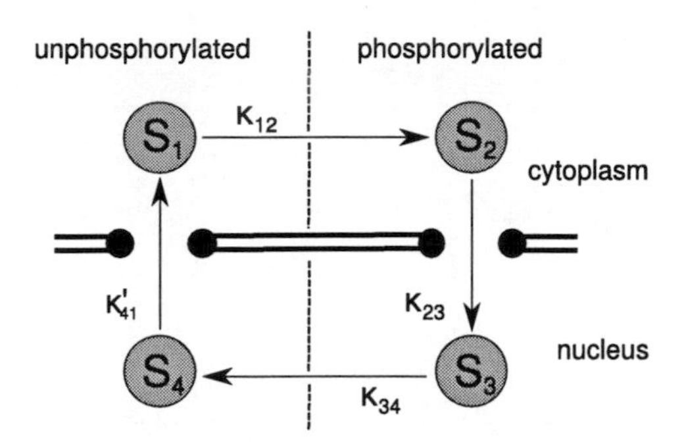

Figure 11.1: Reaction scheme of the hypothetical ΔI transport mutant.

node S_1 in the hypothetical ΔI pathway lacking shuttling evaluates to

$$\widehat{C}_1^4 = C_{12}^4.$$

In the wild-type system the corresponding node concentration control is given by

$$\widehat{C}_1^4 = C_{12}^4 + C_{14}^4 \,,$$

taking into account the additional control of the shuttling process κ_{14}.

The sign rule of Equation (10.30) states that the control of the import process κ_{14} on S_4 must be positive:

$$C_{14}^4 > 0.$$

Since $\widehat{C}_1^4$ has the same value in both systems, it follows immediately that the import step in the wild-type pathway decreases the control of the receptor tyrosine kinase on S_4. Analytical evaluation of C_{14}^4 shows that the control of the Jak kinase is reduced by

$$\Delta C_{12}^4 = C_{14}^4 = \frac{\kappa_{14}}{\kappa_{12} + \kappa_{14}} - \frac{\kappa_{14}}{\kappa_{41}} S_1.$$

By writing the kinase control coefficient as a function of the reaction rates κ_{ij}

$$C_{12}^4 = \frac{\frac{1}{\kappa_{12} + \kappa_{14}} \left(1 + \frac{\kappa_{12}}{\kappa_{23}} + \frac{\kappa_{12}}{\kappa_{34}}\right) - \frac{1}{\kappa_{23}} - \frac{1}{\kappa_{34}}}{\frac{1}{\kappa_{12}} + \frac{1}{\kappa_{23}} + \frac{1}{\kappa_{34}} + \frac{1}{\kappa_{41}} + \frac{\kappa_{14}}{\kappa_{12}\,\kappa_{41}}}, \tag{11.8}$$

one can verify that the kinase control on the unphosphorylated nuclear Stat pool can get arbitrarily small if the shuttling rate κ_{14} is large compared to the remaining transport and reaction rates.

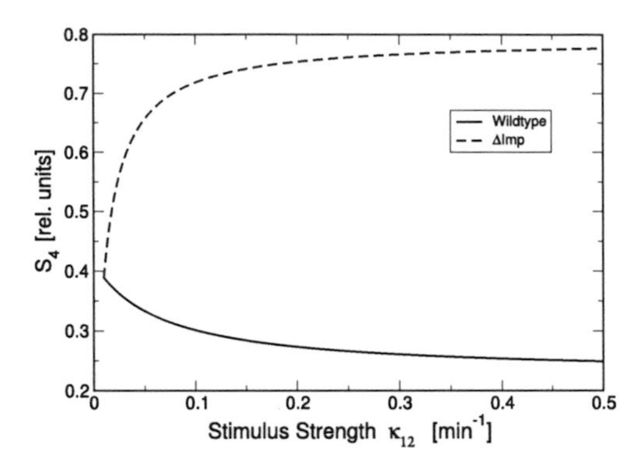

Figure 11.2: Concentration of unphosphorylated nuclear Stat S_4 as function of the stimulus strength κ_{12} in the wild-type and ΔI mutant model. The reduced export rate κ'_{41} of the ΔI mutant was matched at the reference phosphorylation rate $\kappa_{12} = 0.01 \text{ min}^{-1}$.

Due to the decreased control of the receptor-associated kinase on the concentration of S_4, variations in the amount of activated receptor caused by cytokine stimulation of different strengths do not result in pronounced changes of the nuclear concentration of unphosphorylated Stat, as shown by the stimulus-response curve in Figure 11.2. Thus, the shuttling of unphosphorylated molecules stabilises the nuclear concentration of this Stat pool during cytokine induced activation of the Jak/Stat pathway. Interestingly, this behaviour can also be observed in the numerical simulations obtained from the detailed Stat model, as shown in Figure 4.3, where the change of Y_n during cytokine stimulation is relatively small.

Shuttling Increases Response Control of Receptor

Similar to the above analysis of the stabilising effect of the import pathway on the concentration of nuclear unphosphorylated Stat1 we can investigate how nucleo-cytoplasmic shuttling affects the primary pathway response, that is the concentration change of transcriptional active phospho-Stat1 in the nucleus induced by cytokine stimulation of the cell.

In the hypothetical ΔI pathway without nuclear import of unphosphorylated Stat the node concentration control of S_1 on the nuclear phospho-Stat1 dimer pool is determined solely by the control of the kinase step

$$\widehat{C}_1^3 = C_{12}^3.$$

For the wild-type pathway the corresponding expression for the node control is

$$\widehat{C}_1^3 = C_{12}^3 + C_{14}^3,$$

showing that the total control is distributed over the phosphorylation and import process.

Since the import step is not part of any allowed sub-path leading to S_3 (see Chapter 10.1, page 69), it follows from the sign rule of Equation (10.25) that the control of κ_{14} on the nuclear concentration of phosphorylated Stat1 is negative:

$$C_{14}^3 < 0.$$

Analogously to the analysis presented in the previous section, in both signal transduction systems $\widehat{C}_1^3$ must be identical, thus the node concentration control summation rule implies that the shuttling process in the wild-type pathway increases the control of the Jak kinase by

$$\Delta C_{12}^3 = |C_{14}^3| = \frac{\kappa_{14}}{\kappa_{41}}\overline{S}_1.$$

Evaluating C_{12}^3 as a function of the rate constants κ_{ij}

$$C_{12}^3 = \frac{\frac{1}{\kappa_{12}} + \frac{\kappa_{14}}{\kappa_{12}\,\kappa_{41}}}{\frac{1}{\kappa_{12}} + \frac{1}{\kappa_{23}} + \frac{1}{\kappa_{34}} + \frac{1}{\kappa_{41}} + \frac{\kappa_{14}}{\kappa_{12}\,\kappa_{41}}} \tag{11.9}$$

shows that the control approaches its maximal value 1 when the import κ_{14} is made large compared with the other rate constants.

This result reveals that the shuttling of the inactive molecules makes the wild-type Stat pathway more sensitive to variations in receptor activation, increasing the response of the system to small changes in the extracellular ligand concentration. Numerical calculation of a stimulus-response curve for the wild-type and ΔI pathway indeed shows that the wild-type system response curve exhibits a larger slope over the whole range of stimulation strengths, see Figure 11.3. Thus, the wild-type system can respond more sensitive to changes of the extracellular stimulus than the ΔI system with the pure cyclical pathway design.

We can understand this behaviour in a qualitative and more mechanistic way if we consider the concentration changes of the cytoplasmic unphosphorylated Stat fraction induced by increased receptor stimulation. In both systems, an enhanced phosphorylation rate will lead to an initial decrease of the concentration of S_1. In the wild-type pathway, this decrease will also reduce the nuclear import of the unphosphorylated molecules, leading to an enhanced net export of inactive Stat from the nucleus. The enhanced export will stabilise the cytoplasmic concentration in the wild-type system and buffer the drop of S_1 so that the phosphorylation flux in this pathway is larger when compared with the ΔI system.

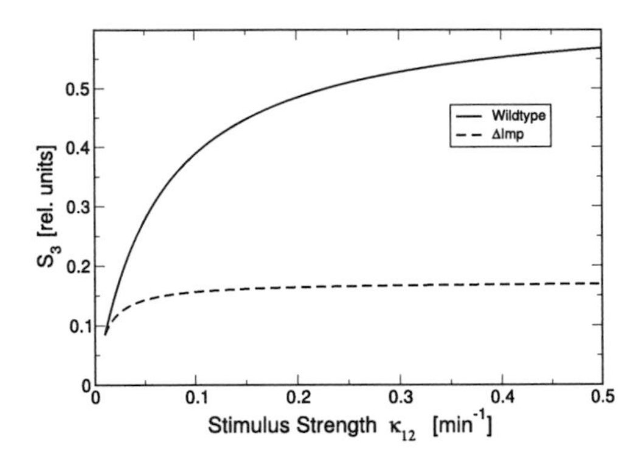

Figure 11.3: Stimulus-Response curve of phosphorylated nuclear Stat S_3 as function of the stimulus strength κ_{12} in the wild-type and ΔI mutant model. The reduced export rate κ'_{41} of the ΔI mutant was matched at the reference phosphorylation rate $\kappa_{12} = 0.01$ min^{-1}.

Sensitivity of Integrated Response is Enhanced by Shuttling

The results presented in the previous section have been derived using the steady state solution of the linearised core model of Stat signalling. The dynamic *in vivo* response of Stat pathways to physiological stimuli is typically transient, compare the experimental data in Figures 4.1 and 4.2. As mentioned already, due to the ample differences of the timescales for activation and deactivation of the pathway, the maximal amplitude of the biological system is well characterised by the steady state solution of the linear Stat model. However, the biological response is not solely determined by the amplitude of the system response. Another important feature is the so called *integrated response* [Heinrich et al., 2002], which connects the amplitude of the system's response with its duration, yielding the total amount of activated transcription factor produced during the stimulation period. Mathematically, the integrated response of the Stat system is defined as the integral of nuclear phosphorylated fraction $P_n(t) = Y_n^p(t) + Z_s(t) + Z_u(t)$ over time:

$$I_S = \int_{t_0}^{\infty} P_n(t)dt, \tag{11.10}$$

where t_0 denotes the start of stimulation of the pathway. That the integrated response is indeed a biological relevant property for transcriptional activation has been shown by experimental measurements for the MAP kinase pathway [Asthagiri et al., 2000].

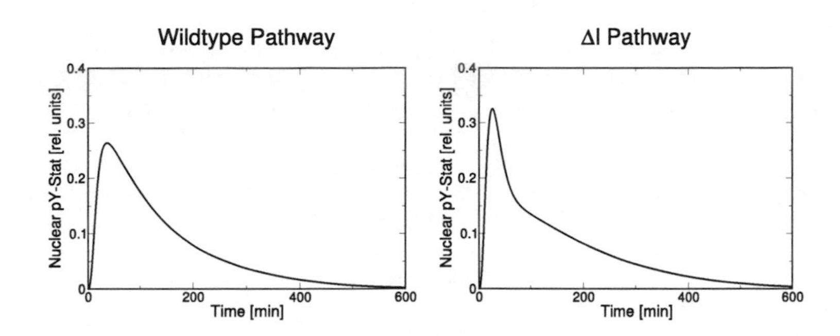

Figure 11.4: Numerical simulation of the nuclear concentration of activated Stat1 during a typical cytokine stimulation experiment for the wild-type and the hypothetical ΔI pathway. Both systems exhibit the same integrated response (area under the curves of the nuclear phosphorylation signal).

To quantify the effect of the shuttling process on the integrated response of the transient output of the Stat1 signalling network, we applied the results of the previous sections to the detailed Jak/Stat1 model, comparing the wild-type system with a pathway design lacking the import process. The wild-type model Equations (3.1)–(3.10) and the according parameters in Table 3.4 were chosen as a reference system and the integrated response was calculated for the concentration of active Stat1 in the nucleus. Then a model of a hypothetical pathway ΔI lacking the import process (k_9=0) was derived from the detailed model, where the export rate k_8 was reduced so that both systems exhibited the same integrated response. Figure 11.4 shows the time-course of phosphorylated Stat1 in the nucleus as simulated by both models. The dataset derived from the ΔI model exhibits a marked peak at the begin of the activation profile, which is caused by the instantaneous phosphorylation of all Stat molecules which are located exclusively in the cytoplasm prior to stimulation. Due to this phosphorylation burst, the export rate k_9 has to be lowered to $\approx$ 10% of the reference model value to maintain the same integrated response in both systems.

The differential control on the system's integrated response by reaction and transport parameters and varying external stimuli was tested for the two pathway designs. Analogously to the control analysis presented in Chapter 5.1, a single intrinsic or extrinsic model parameter was varied between 25% and 400% of its reference value while the remaining parameters were held constant at their specific reference value. Then the integrated response was calculated numerically for both models and compared with its value in the system's reference state. Except for two parameters the response characteristics of both models were nearly identical. Interestingly, profound differences of the control over the integrated response in the wild-type and ΔI pathway were found for the parameters R_0 and k_4, which represent the total amount of IFN-γ receptors and the on-rate constant for the binding of Stat1 to the receptor, respectively. In the detailed pathway models the product of these two parameters characterises the recruitment of Stat1 to the receptor, where the transcription factor eventually gets phosphorylated by the Jak kinase. In both models the kinase reaction itself (denoted by the reaction rate k_5) exhibits almost no control on the integrated response since the recruitment of Stat1 to the receptor is a limiting step for the whole phosphorylation sequence, as we have already shown in the response control analysis in Section 5.1.1.

Figure 11.5 shows the plots of the response control of R_0 and k_4 in both pathways. The plots show that in the wild-type pathway the sensitivity to negative changes of receptor-related processes is about 30% higher as in the ΔI pathway, while for positive variations the sensitivity is nearly doubled. Similar results are found if the numerical control coefficient itself, which is equivalent to the slope of the curves at the origin, is compared: The coefficients of the ΔI pathway for k_4 and R_0 are 0.251 and 0.231, respectively, while for the wild-type system the corresponding values are 0.404 and 0.389. Interestingly the differences of the control coefficients in both systems correspond to the numerical control coefficient of the import process in the wild-type system, which is -0.176. Thus, these numerical results obtained from the detailed Stat model confirm the analytical results derived from the simplified core model and show that the control of the stimulus over the time-integrated response in the wild-type system is strongly enhanced by the shuttling of transcriptionally inactive Stat1.

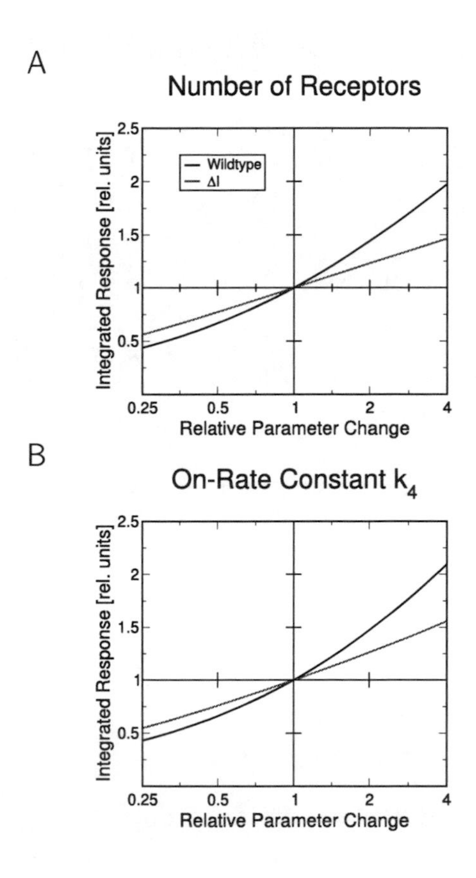

Figure 11.5: Sensitivity of the wild-type and ΔI pathways to variations in receptor-related parameters. Shown is the relative change of the integrated response of the nuclear phosphorylated Stat1 fraction as a function of the change in the number of activateable interferon-γ receptors R_0 (Figure A) and in the on-rate k_4 of Stat1 binding to the receptor (Figure B) for the ΔI (red) and wild-type model (black).

12

Discussion of Part II

State Lifetimes and Steady State Concentration

In part II of this thesis we have developed a linear core model of Stat signal transduction by reducing the mathematical complexity of the detailed model introduced in the first part of this work. After applying rapid-equilibrium approximations for the binding processes of Stat to the cytokine receptor and to DNA the reduced model brings out clearly the linear cyclical design of the Jak/Stat pathway. Interestingly, this cyclical design leads to a dependence between the average cycling time T_C of the Stat signalling proteins in the pathway, the lifetime τ_i of a specific Stat fraction and the steady-state concentration $\overline{S}_i$ of this fraction. We could show the universal relation

$$\overline{S}_i = \frac{\tau_i}{T_C},$$

for all subcellular fractions $\overline{S}_i$ in the Stat signal transduction pathway. This general relationship should be easily applicable to other cyclic signal transduction pathways and similar dynamic systems as well.

Steady State Concentration Determines Network Control

The treatment of signal transduction pathways as networks of linear state transitions uncovered a relation between the steady state concentration of molecules in a specific state m and the total control of all the transition steps emanating from this state. Interestingly, the total control exerted on any other state $i \neq m$ in the network is equal to the steady-state fraction $\overline{X_m}$. Thus, if only a single process emanates from the node corresponding to the state m, the control of this process simply equals $\overline{X_m}$. However, when several steps emanate from node m, some (or all) of them can have high control even when $\overline{X_m}$ is small. This is the case when there are control coefficients with different signs, that can partially cancel each other in the summation in Equation (10.14).

The state transition approach enabled us to derive rules determining the sign and magnitude of control of specific network processes. The application of these principles to the Jak/Stat system showed that due to the network design, the nuclear dephosphorylation reaction always has the strongest control over the pathway response, independent of specific values of the model parameters.

Competing Processes Change Control Distribution

Examination of the shuttling process of inactive molecules in the Jak/Stat system revealed that the nuclear import of unphosphorylated Stat1 changes the control of the phosphorylation reaction with respect to two different aspects. On the one hand, the control of the phosphorylation reaction over the nuclear concentration of phosphorylated Stat1 is amplified, leading to an increased stimulus sensitivity of the system and enabling the cell to sense differences in receptor activation more accurately. On the other hand, the same import process also leads to a decreased control of the phosphorylation step over the nuclear concentration of unphosphorylated Stat1, resulting in homeostasis of this Stat fraction. Since unphosphorylated Stat molecules have been reported to play an essential role in the regulation of several cytokine-independent genes [Chatterjee-Kishore et al., 2000, Yang et al., 2005], a relative robustness of the nuclear unphosphorylated Stat pool against varying cytokine stimulation strengths of the cytokine receptors might be essential to avoid crosstalk between different extracellular signals. A tightly balanced interplay of the export and import pathways might help the cell to decouple the regulation of different target gene groups. Interestingly a similar cyclic pathway design together with the shuttling of inactive transcription factors has also been found in other signal transduction systems, for example for Smad signalling [Inman et al., 2002, Xu and Massague, 2004]. Thus, this pathway topology pattern might represent a evolutionary developed network design. The changed network control properties caused by the constitutive shuttling of inactive Stat highlight an interesting biological network design principle: The introduction of a competing process at one node can change the control distribution in the network significantly. It remains an interesting area for further study whether the general approaches presented in the last chapters can be extended to study the control distribution in more complex signalling networks.

Part III

Modelling Stat1 Nuclear Transport Mutants

13

Regulatory Potential of Nucleo-Cytoplasmic Transport

It is well-known that Jak/Stat signalling networks are regulated at multiple levels and by different mechanisms. For example, the pancellular Stat phosphorylation level is determined by the activation and inactivation of Stat1 proteins by opposing kinases and phosphatases. Other regulatory targets in the pathway are DNA binding and the recruitment of transcriptional co-activators. In contrast, the regulatory potential that is inherent to the dynamic redistribution of the shuttling signal transducer between the nuclear and cytoplasmic compartment has gained relatively little attention.

The phosphorylation state of a signalling protein is generally determined by the balance of the activities of kinases and phosphatases. However, the continuous dynamic redistribution of Stat1 between the cytoplasmic and nuclear compartments together with the spatial containment of the Jak kinase and the main phosphatase activity to the cytoplasm and the nucleus, respectively, implies that the regulation of Stat1 phosphorylation cannot be understood solely in terms of the activities of Jak kinases and Stat1-specific phosphatases. Rather the nuclear translocation determining the access of Stat1 to its kinases and phosphatases should play a critical role as well.

Distinct Role of Import and Export

The theoretical analysis of the Stat1 signalling pathway using the detailed Jak/Stat1 model in part I of this work and the more general analysis of cyclic signalling pathways with the Stat core model in part II have both revealed a distinct role of the nucleo-cytoplasmic transport processes in the signalling pathway. The shuttling of the unphosphorylated molecule is expected to decouple the nuclear concentrations of the unphosphorylated and phosphorylated Stat fractions and should increase the pathway response.

Furthermore, the detailed Stat1 model has predicted that the nuclear export of unphosphorylated Stat1 can be a limiting step for the nuclear accumulation of pY-Stat1, whereas the import of the phosphorylated protein into the nucleus does not exert rate limitation. This may appear surprising at first, because anti-body trapping and microinjection experiments have shown that both steps have very similar velocity (compare Chapter 3.2.1). However, the net export rate of inactive Stat1 from the nucleus is lowered dramatically by the rapid re-import. This implies that tilting the balance of the shuttling in favour of nuclear import - either by increasing the import rate or decreasing the export rate of the unphosphorylated molecule - should strongly downregulate rephosphorylation. Similarly, increasing the net export rate should enhance the phosphorylation amplitude. By contrast, interfering with the import rate of the phosphorylated Stat1 molecules should have little effect on the phosphorylation amplitude.

To examine this theoretical predictions, we analysed experimental data from a series of Stat1 nuclear translocation mutants measured by the research group of U. VINKEMEIER. By site-specific mutagenesis the transport rates of the import and export of the phosphorylated and unphosphorylated molecules could be enhanced and the resulting phenotypes could be compared to the predictions of the Stat1 model.

14 Nuclear Import Mutant

We first examined Stat1 transport mutants in which the nuclear import was enhanced by adding nuclear localisation sequences (NLS) of different strengths in between the Stat1 C terminus and a GFP domain. In one case, an attenuated NLS from the large T antigen of SV40 was used, giving rise to a weakly import-enhanced Stat1 mutant (wNLS mutant). In a second construct, a tandem arrangement of the native, non-attenuated NLS from the large T antigen of SV40 was added, yielding a stronger import acceleration (sNLS mutant). As a control, a third mutant was generated by permutating the NLS sequences of the sNLS mutant as to disrupt the physiological NLS function, which we refer to as LSN mutant.

14.1 NLS Mutant Phenotype

The different strengths of the various NLS constructs become apparent by assaying the subcellular Stat1 distribution using fluorescence microscopy (see Figure 14.1). In contrast to wild-type Stat1 displaying a nearly pancellular resting distribution, the wNLS mutant shows a preferentially nuclear localisation already prior to stimulation, with a small remaining cytoplasmic fraction. The sNLS mutant with the stronger NLS exhibits an exclusively nuclear localisation in resting cells. That these differences to the wild-type are due to the added NLS function is confirmed by the resting distribution of the permutation mutant LSN, which is identical to that of wild-type Stat1. Thus, the wNLS and sNLS mutants allow to evaluate the effects of a graded increase in nuclear import rate on Stat1 activation.

Upon stimulation with IFN-γ, the wNLS and LSN mutants accumulate in the nucleus to the same degree as wild-type Stat1, while the already strong nuclear accumulation of the sNLS mutant is not further augmented by interferon treatment. However, both wNLS and sNLS mutants display significant levels of phosphorylation, although they were lower than the wild-type signal (Figure 14.2, solid lines).

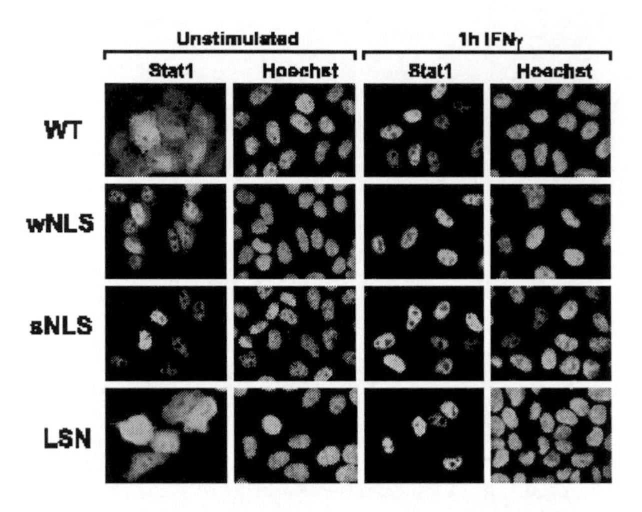

Figure 14.1: Effect of enhanced import of Stat1 on the subcellular distribution of the transcription factor. The addition of NLS sequences of different strengths to Stat1-GFP resulted in a gradual increase of nuclear Stat1 in resting cells and unaltered nuclear accumulation upon cytokine stimulation. Shown are fluorescence micrographs of resting and IFN-γ-stimulated (1 h) cells transiently expressing wild-type Stat1-GFP (WT), the weak import mutant (wNLS), the strong import mutant (sNLS) and control mutant protein (LSN) with a permutated nuclear location sequence. The cells were fixed, stained for DNA (Hoechst) and processed for the detection of GFP fluorescence. (Experimental data: T. MEYER, FMP Berlin)

Remarkably, even the sNLS mutant protein, which appears to be fully nuclear already in resting cells, becomes phosphorylated. This observation indicates that, like the wild-type, both mutant proteins cycle between the cytoplasm and the nucleus and thus have access to the IFN-γ receptor/Jak complex at the plasma membrane. That phosphorylation is continuously maintained is corroborated by the rapid decline of phospho-Stat1 caused by the application of the kinase inhibitor staurosporine (Figure 14.2, dashed lines).

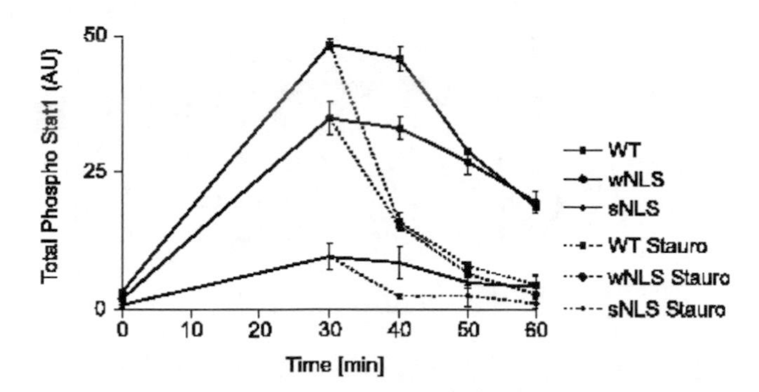

Figure 14.2: Both Stat1 import mutants wNLS and sNLS exhibit detectable phosphorylation levels upon IFN-γ stimulation and are sensitive to the kinase inhibitor staurosporine. Shown is the quantification of Western blot experiments with whole-cell extracts from U3A cells expressing either wild-type Stat1-GFP (WT) or the import mutants wNLS and sNLS. Cells were incubated with 5 ng/ml IFN-γ (solid lines). For the kinase inhibition experiment, 500 nM staurosporine was added after $t = 30$ min (dashed lines). (Experimental data: T. MEYER, FMP Berlin)

14.2 NLS Mutant Model

We have simulated the mutant phenotypes in the detailed Jak/Stat1 model by augmenting the nuclear import rate constants for both phospho-Stat1 dimers and the unphosphorylated protein (k_8 and k_{13}, respectively) by the same value, compare the schematic view shown in Figure 14.3. For the weak import mutant, we choose $k_8 = 0.43\,\text{min}^{-1}$ and $k_{13} = 0.26\,\text{min}^{-1}$, thereby doubling the import of the unphosphorylated molecule. For the sNLS mutant, the rates are set to $k_8 = 2.77\,\text{min}^{-1}$ and $k_{13} = 2.6\,\text{min}^{-1}$, which accelerates the import of Y-Stat1 by a factor of 20.

The nuclear presence of the protein under resting and stimulated conditions increases with the degree of import acceleration (Figure 14.4 A), while the phosphorylation levels decrease (Figure 14.4 B). These simulation closely match the corresponding experimental data shown in Figures 14.1 and 14.2.

The decrease of the phosphorylation signal is solely due to the accelerated import of the unphosphorylated protein (k_{13}), while the increased import rate of the phospho-dimer (k_8) has no effect on the system behaviour. This can be demonstrated by a numerical simulation in which only the import rate of the phospho-dimer k_8, but not of the unphosphorylated protein, is augmented. The simulation

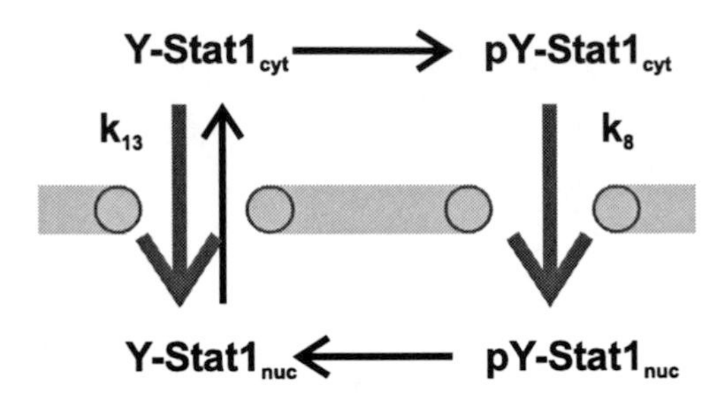

Figure 14.3: Schematic view of the import acceleration in the model of the wNLS and sNLS transport mutants. Both the rate constant for the import of unphosphorylated Stat1 (k_{13}) as well as the import rate for the phosphorylated protein (k_8) are accelerated by the same value.

of such a hypothetical import mutant yields practically the same behaviour as the wild-type protein both with respect to nuclear accumulation and phosphorylation (Figures 14.4 C and D). Conversely, a selective augmentation of the import rate of the unphosphorylated protein k_{13} reproduces the experimental data for the wNLS and sNLS import mutants. The model simulations are then practically identical to the curves shown in Figures 14.4 A and B for import augmentation of both unphosphorylated and phosphorylated proteins (results not shown).

Taken together, the experimental data and the according model simulations show that the rephosphorylation rate of the unphosphorylated Stat1 protein can be strongly regulated by the velocity of its nuclear import. This is the case because the rapid nuclear import of unphosphorylated Stat1 competes with its activation by the Jak tyrosine kinases in the cytoplasmic compartment.

Import of Y-Stat1 and pY-Stat1 accelerated:

A

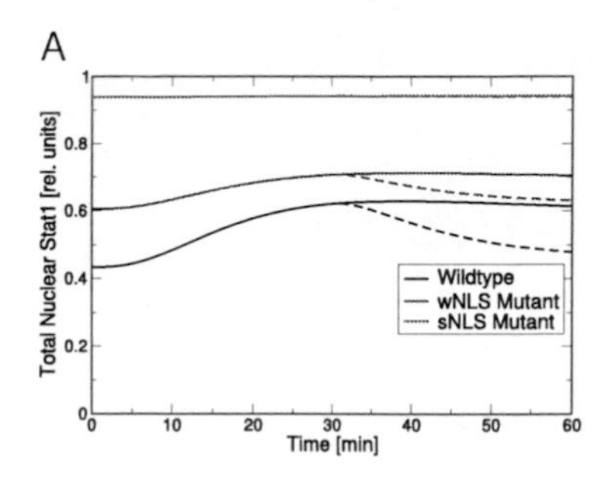

B

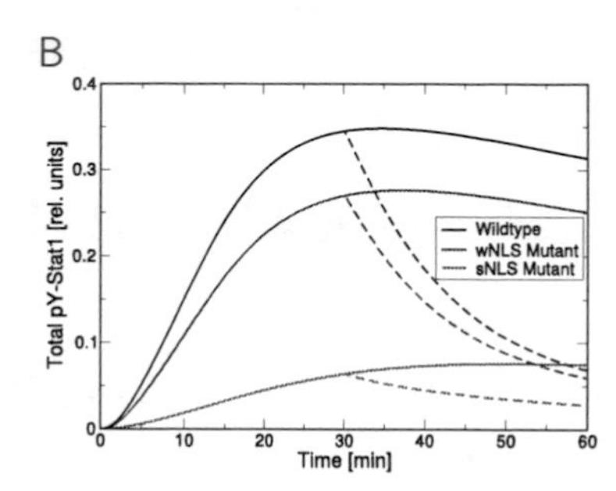

Only Import of pY-Stat1 accelerated:

C

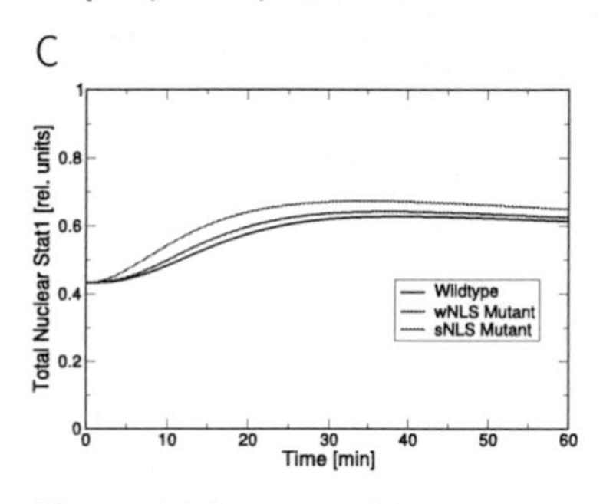

D

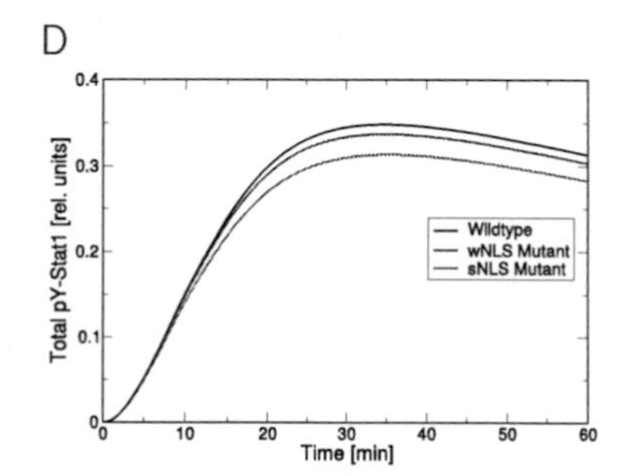

Figure 14.4: **A**: Model simulation of the total nuclear concentration of Stat1 for wild-type (black line), weak import mutant (blue line) and the strong import mutant (red line). The graded increase of the resting concentration is visible at $t = 0$ min. Following the application of an IFN-γ stimulus for $t > 0$ min, the increase of the nuclear concentration for the wild-type and the wNLS mutant, as well as the unchanged nuclear accumulation of the sNLS mutant closely match the experimental data (compare Figure 14.1). The dashed lines show the model simulation of the kinase inhibition experiments, where the inhibitor staurosporine was added at $t = 30$ min. **B**: Simulation of the phosphorylation level during IFN-γ stimulation for wild-type, wNLS and sNLS mutant (solid lines: continuous Jak kinase activity, dashed lines: Jak kinase activity blocked after $t = 30$ min). Import acceleration leads to a graded decrease in Stat1 phosphorylation, as shown by the experimental data in Figure 14.2. **C** and **D**: Simulations of an import mutant where only the nuclear import of phosphorylated Stat1 (k_8) is augmented. Shown are the model results of identical simulation protocols as in **A** and **B**, respectively.

15

Nuclear Export Mutant

Next we asked how the acceleration of nuclear export affects Stat1 activation. In the wild-type system, the nuclear export is restricted to the unphosphorylated form of the protein. To enhance this nuclear export pathway, a Stat1 mutant was generated in the lab of U. VINKEMEIER by adding the nuclear export sequence (NES) located in the DNA-binding domain of Stat1 heterotopically to the C-terminus of the protein (Stat1-NES). The NES export signal is a recognition sequence for the Crm1 export receptor and strongly increases the Crm1-dependent export.

15.1 NES Mutant Phenotype

The subcellular distribution of the Stat1-NES transport mutant was measured using fluorescence microscopy and compared to wild-type Stat1 (Figure 15.1). Resting cells showed a nearly complete cytoplasmic localisation of Stat1-NES, which contrasts with the pancellular distribution of wild-type Stat1 (first row in Figure 15.1, 'unstim.'). Treatment with the nuclear export inhibitor leptomycin B (LMB), which blocks the Crm1-specific export pathway, but not the energy-independent shuttling of Y-Stat1 via the NPC-dependent pathway, resulted in the pancellular distribution of Stat1-NES comparable with the wild-type Stat1 (Stat1-WT) distribution in resting cells ('+LMB'). Therefore, unphosphorylated Stat1-NES, like the wild-type protein, is continuously imported into the nucleus. However, the enhanced export rapidly clears the nucleus from unphosphorylated Stat1-NES. Stimulation by interferon-γ did not result in nuclear accumulation of the signalling molecule, in contrast to Stat1-WT ('+IFN-γ'). This is not due to a phosphorylation defect, as Stat1-NES becomes strongly tyrosine-phosphorylated (see below). Nuclear accumulation on IFN-γ stimulation was restored in Stat1-NES when nuclear export was blocked by the simultaneous addition of LMB ('+LMB/+IFN-γ'). These data show that phospho-Stat1-NES is imported into the nucleus.

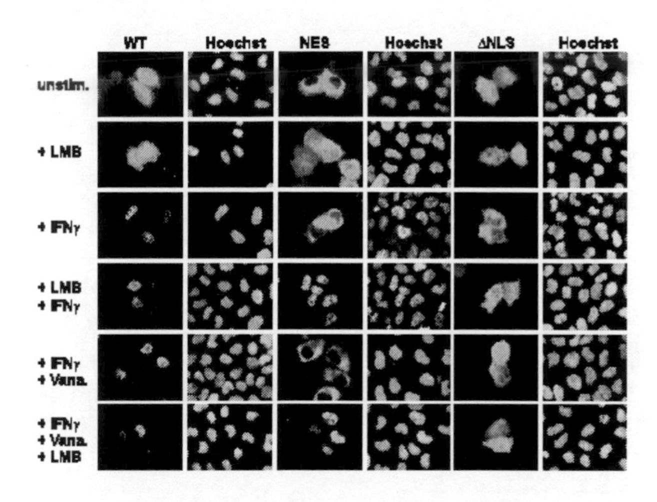

Figure 15.1: Effect of enhanced nuclear export of phosphorylated and unphosphorylated Stat1 on subcellular distribution and cytokine-induced phosphorylation. Shown are the nuclear accumulation behaviour of wild-type Stat1, the export mutant NES, and the import mutant ΔNLS in HeLa cells and the corresponding Hoechst-stained nuclei. The cells were either left untreated (unstim.; first row) or incubated for 1.5 h with 10 ng/ml of the export inhibitor leptomycin B (+LMB; second row), 5 ng/ml of interferon (IFNγ; third row), or a combination of both (+LMB +IFNγ, fourth row). Additionally, cells were treated for 1 h with 5 ng/ml IFN-γ followed by a 30 min incubation in either 5 ng/ml IFN-γ and 0.8 mM vanadate (+IFNγ +Vana.; fifth row) or IFN-γ, vanadate and LMB (+IFNγ +LMB; sixth row). The cells were fixed, stained for DNA (Hoechst) and processed for the detection of GFP fluorescence. (Experimental data: T. MEYER, FMP Berlin)

To demonstrate that the lack of nuclear accumulation in the absence of LMB is due to rapid export of the phospho-protein, we stimulated the cells with IFN-γ and at the same time blocked dephosphorylation by vanadate. Under this condition, phospho-Stat1-NES did not accumulate in the nucleus, whereas Stat1-WT was trapped there ('+IFN-γ/+Vana.'). Only the additional inhibition of export by LMB rescued the defective nuclear accumulation of Stat1-NES ('+IFN-γ/+Vana./+LMB'). Taken together, these data demonstrate that phospho-Stat1-NES enters the nucleus, but cannot be retained there due to its rapid export as a phospho-protein. This conclusion is further corroborated by the comparison with

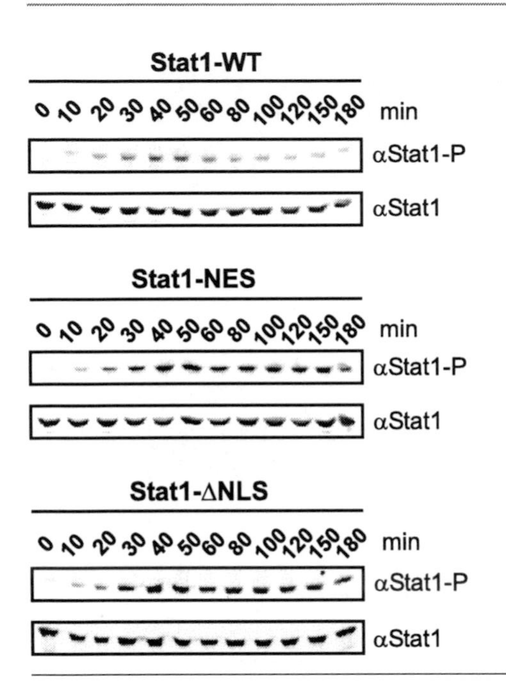

Figure 15.2: Prolonged and enhanced phosphorylation of both Stat1-NES and Stat1-ΔNLS as compared to the wild-type protein. U3A cells expressing the respective Stat1-GFP were stimulated with IFN-γ for the indicated times. Shown are Western blots from whole-cell extracts developed with a Stat1-specific phospho-tyrosine antibody and reprobed with a Stat1 antibody. (Experimental data: T. MEYER, FMP Berlin)

the Stat1-ΔNLS mutant (described in Chapter 3.2.2), which is not imported into the nucleus as a phospho-protein. Consequently, its nuclear accumulation cannot be induced by any of the protocols used (column 5 in Figure 15.1).

Treating cells with IFN-γ resulted in a higher and prolonged phosphorylation of Stat1-NES compared to the wild-type (Figure 15.2). This is a direct consequence of the export acceleration, leading to a predominant localisation of the mutant protein in the cytoplasmic compartment. In the cytoplasm the molecules are exposed directly to the strong phosphorylation activity of the Jak kinase and additionally protected from the strong nuclear phosphatase TC45, only subjected to the comparatively weak cytoplasmic phosphatase. This conclusion is confirmed by the fact that the Stat1-ΔNLS mutant with a completely different phenotype, for which the phospho-protein does not enter the nucleus, exhibits a similarly augmented phosphorylation kinetics.

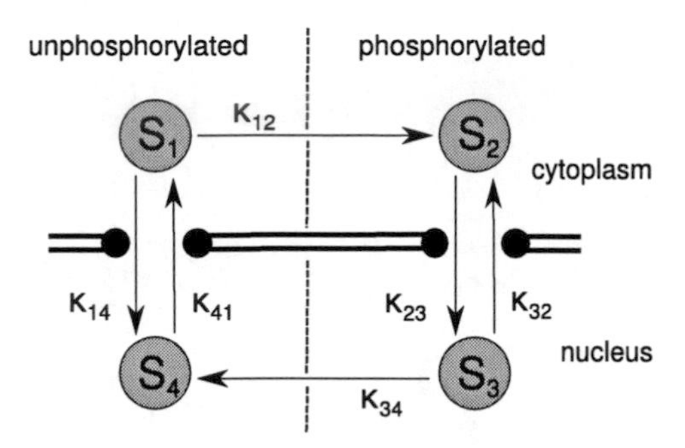

Figure 15.3: Scheme of the Stat-NES core model with disrupted network structure due to the nuclear export κ_{32} of the phosphorylated protein S_3.

15.2 Structural Analysis of the NES Pathway

The experimental data shows that the presence of an additional export sequence in the Stat1-NES mutant causes a phenotype for which the cyclical structure of the network is disrupted. Unlike the wild-type, Stat1-NES can be exported from the nucleus directly as phospho-protein, without prior dephosphorylation by a nuclear phosphatase. In order to evaluate the consequence of this short-circuited network structure, we first compared the wild-type and the NES mutant network using the analytic solutions of the simplified core model of Stat signalling introduced in part II of this work. A schematic view of the NES mutant core model network is shown in Figure 15.3 (compare with Figure 8.1).

Adding a nuclear export process κ_{32} for the phosphorylated protein to the Stat core model (8.10)–(8.13) and solving the equation system for its steady state solution yields

$$
\begin{aligned}
\overline{S_1^n} &= \frac{1}{T}\frac{1}{\kappa_{12}} \\
\overline{S_2^n} &= \frac{1}{T}\frac{1}{\kappa_{23}}\left(1+\frac{\kappa_{32}}{\kappa_{34}}\right) \\
\overline{S_3^n} &= \frac{1}{T}\frac{1}{\kappa_{34}} \\
\overline{S_4^n} &= \frac{1}{T}\frac{1}{\kappa_{41}}\left(1+\frac{\kappa_{14}}{\kappa_{12}}\right)
\end{aligned}
$$

with

$$T = \frac{1}{\kappa_{12}} + \frac{1}{\kappa_{23}}\left(1 + \frac{\kappa_{32}}{\kappa_{34}}\right) + \frac{1}{\kappa_{34}} + \frac{1}{\kappa_{41}}\left(1 + \frac{\kappa_{14}}{\kappa_{12}}\right).$$

Export Block Enhances System Response

To simplify the analysis we assume that the import and export processes of the phosphorylated and unphosphorylated molecules proceed with the same velocity in each direction:

$$\kappa_{32} = \kappa_{23} \quad \text{and} \quad \kappa_{14} = \kappa_{41}. \tag{15.1}$$

Then we have for the steady state concentration of the transcriptionally active Stat protein $\overline{S}_3^n$ in the nucleus:

$$\overline{S}_3^n = \frac{\frac{1}{\kappa_{34}}}{\frac{2}{\kappa_{12}} + \frac{1}{\kappa_{23}} + \frac{2}{\kappa_{34}} + \frac{1}{\kappa_{41}}}. \tag{15.2}$$

Assuming fast transport processes

$$\kappa_{23} \to \infty \quad \text{and} \quad \kappa_{41} \to \infty, \tag{15.3}$$

equation (15.2) simplifies to

$$\overline{S}_3^n = \frac{\frac{1}{\kappa_{34}}}{\frac{2}{\kappa_{12}} + \frac{2}{\kappa_{34}}} \tag{15.4}$$

Applying assumptions (15.1) and (15.3) to the wild-type steady state Equation (9.1) we get for the concentration of nuclear pY-Stat1

$$\overline{S}_3^w = \frac{\frac{1}{\kappa_{34}}}{\frac{2}{\kappa_{12}} + \frac{1}{\kappa_{34}}}. \tag{15.5}$$

Analysis of the steady state equations (15.5) and (15.4) reveals that the wild-type system exhibits a stronger nuclear phosphorylation response than the NES mutant if both systems are exposed to the same stimulus. In the limit of very strong stimulation strengths $k_{12} \to \infty$, all Stat molecules are phosphorylated and transported into the nuclear compartment in the wild-type pathway. Equation (15.4) shows that the NES mutant can only reach a maximal nuclear phosphorylation of 50%:

$$\overline{S}_{3,\max}^w = 1 \quad \text{and} \quad \overline{S}_{3,\max}^n = \frac{1}{2}. \tag{15.6}$$

If we have for the phosphorylation rate κ_{12}^w in the wild-type pathway

$$\kappa_{12}^w < 2\kappa_{34},$$

the nuclear phosphorylation will be below 50% and the NES mutant can reach the same phosphorylation level as the wild-type when its phosphorylation rate constant κ_{12}^n is increased by stronger cytokine stimulation according to

$$\kappa_{12}^n = \kappa_{12}^w \frac{2\kappa_{34}}{2\kappa_{34} - \kappa_{12}^w}. \tag{15.7}$$

For example, to reach the experimentally observed phosphorylation level of $\approx$ 33% in the simplified core models, a two times higher number of active IFN-γ receptor/Jak complexes would be necessary in the NES mutant.

Export Block Reduces Deactivation Time

Due to the very different dephosphorylation activities in the cytoplasmic and nuclear compartment, the inclusion of the phosphorylated Stat protein to the nucleus in the wild-type system will also have a distinct effect on the dephosphorylation kinetics of the network. Particularly, the time for deactivation of the signalling pathway by dephosphorylation of the active Stat1 molecules will be affected by the additional export pathway of the NES mutant.

Using the core model, we can estimate the deactivation time of the system as the lifetime of the nuclear phosphorylated Stat fraction after an instantaneous block of the kinase activity at $t = t_{\text{off}}$ by setting $\kappa_{12} = 0$. Then we have for the wild-type pathway

$$\tau_d^w = \frac{1}{\overline{S_3^w}} \int_{t_{\text{off}}}^{\infty} S_3^w(t)\, dt = \frac{1}{\kappa_{23}} + \frac{1}{\kappa_{34}}, \tag{15.8}$$

while the deactivation time in the NES mutant evaluates to

$$\tau_d^n = \frac{1}{\overline{S_3^n}} \int_{t_{\text{off}}}^{\infty} S_3^n(t)\, dt = \frac{1}{\kappa_{23}} + \frac{2}{\kappa_{34}}, \tag{15.9}$$

assuming equal rates for the import and export of the phosphorylated Stat molecules ($\kappa_{32} = \kappa_{23}$).

For a fast import of the phosphorylated protein, the signalling molecules in the NES pathway will take twice as long to be dephosphorylated and deactivated when compared with the wild-type system:

$$\tau_d^n = 2\tau_d^w.$$

Interestingly, these results are independent of the stimulus strength κ_{12}.

The above comparison of the pathway deactivation time and the network response of the Stat1-NES mutant and the wild-type signalling pathway reveals that the

	WT	NES-U	NES-P	NES
total pY-Stat1	0.30	0.40	0.38	0.49
nuclear pY-Stat1	0.25	0.32	0.17	0.21

Table 15.1: Maximal total and nuclear Stat1 phosphorylation levels in the wild-type pathway (WT), the NES transport mutant and the two hypothetical mutants NES-U and NES-P.

export block for the phosphorylated Stat protein in the wild-type system might be a critical design feature of the Jak/Stat networks and changes the characteristics of the system response.

15.3 Detailed Stat1-NES Model

To analyse the measured experimental data of the NES mutant protein in more detail, we simulated the Stat1-NES mutant numerically by adding an export pathway for the phospho-protein (k_{14}) to the detailed Jak/Stat1 model. Additionally we increased the export rate of the unphosphorylated protein k_{13} by the same value as the k_{14} rate constant (compare Figure 15.4). To recover the observed nucleo-cytoplasmic resting distribution of approximately 15 : 85, the additional export was set to $k_{14} = 0.5\,\text{min}^{-1}$.

Both experimentally observed lack of nuclear accumulation and the higher phosphorylation amplitude were recovered in this simulation (Figures 15.4 B and C). Crucially, the amount of nuclear phospho-Stat1-NES after interferon-γ stimulation is predicted to be lower than the wild-type level of $\approx 25\%$, but still significant (Figure 15.4 C, dashed line). Blocking the export of the phospho-protein to mimic the effect of the export inhibitor LMB reverts the phenotype completely back to the wild-type. The predicted dramatic change of the subcellular distribution of total and phosphorylated Stat1-NES upon LMB-induced export inhibition is supported by additional immunofluorescence studies (Figure 15.5).

To evaluate the specific contribution of the two independent export pathways of phosphorylated and unphosphorylated Stat1 on the changed pancellular and nuclear phosphorylation level in the mutant, we compared the NES system to two hypothetical transport mutants. In the NES-U pathway, only the export of the unphosphorylated molecules is accelerated and k_{13} is set to the same rate as in the NES model. For the NES-P pathway, we did not change the transport of the unphosphorylated protein, but introduced nuclear export of phosphorylated Stat1 with an identical rate constant $k_{14} = 0.5\ \text{min}^{-1}$ as in the NES mutant model. Table 15.1 lists the maximal total and nuclear phosphorylation levels of

A

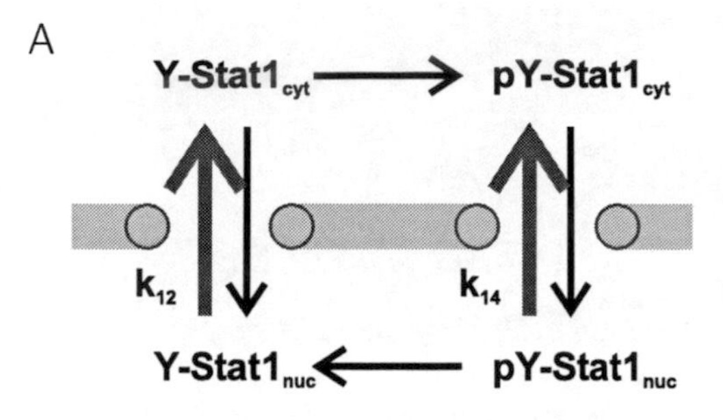

B

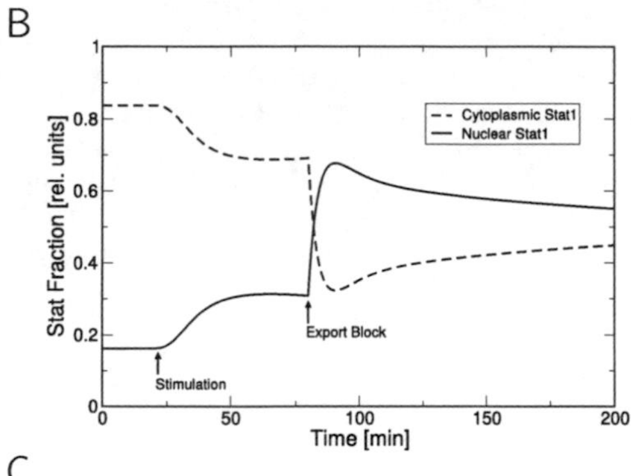

C

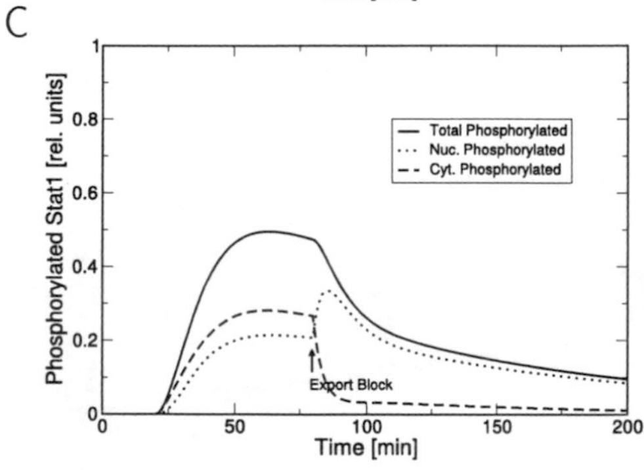

Figure 15.4:
A: Schematic view of the additional nuclear export pathway for phosphorylated Stat1 (κ_{14}) and the export acceleration of unphosphorylated Stat1 (κ_{12}) in the model of the Stat1-NES mutant.
B: Simulation of the NES mutant IFN-γ-LMB experiments shown in Figure 15.1. Plotted is the nucleo-cytoplasmic distribution of Stat1-NES prior to stimulation ($t < 20$ min), throughout IFN-γ stimulation ($20\,\text{min} < t < 80\,\text{min}$) and after addition of the export inhibitor LMB ($t > 80\,\text{min}$).
C: Phosphorylation kinetics of Stat1-NES according to the experimental protocol in **B**. Shown are the total cellular phosphorylation level (solid line) and the cytoplasmic (dashed line) and nuclear (dotted line) phosphorylated fractions. Compare the experimental data in Figure 15.2.

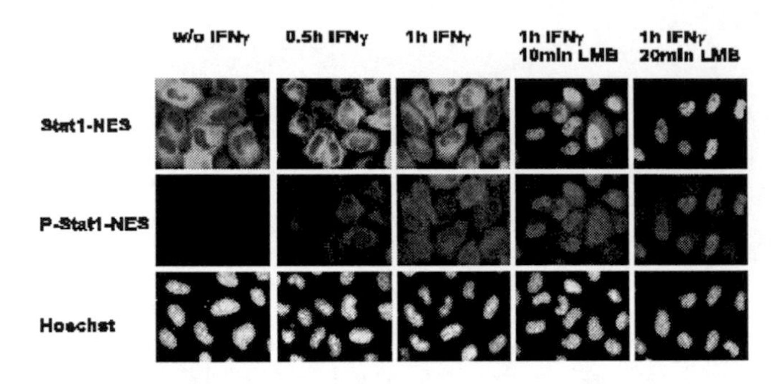

Figure 15.5: Upon blocking the export receptor CRM1 by LMB phosphorylated Stat1-NES is rapidly retained in the nucleus of IFN-γ stimulated cells. Shown is the time-course of Stat1-NES-GFP intracellular redistribution and phosphorylation in HeLa cells before and after stimulation with IFN-γ (30 min and 1h, respectively) as well as in IFN-γ-pretreated cells (1h) incubated for 10 min or 20 min with LMB. The subcellular Stat1 distribution was visualised microscopically by directly tracing the GFP fluorescence while phospho-Stat1 was detected by means of immunocytochemistry. The positions of the corresponding Hoechst-stained nuclei are depicted in the lower row (Experimental data: T. MEYER, FMP Berlin)

the NES, NES-U, NES-P and of the wild-type pathway. The data reveals the two independent effects of the export of phosphorylated and unphosphorylated Stat1 on the pancellular phosphorylation signal: The acceleration of the export of unphosphorylated Y-Stat1 increases the phosphorylation by the cytoplasmic Jak kinase by 10% (NES-U), while the export of the phosphorylated Stat1 increases the cytoplasmic concentration of pY-Stat1 and protects the phosphorylated protein from the strong nuclear phosphatase activity, leading to an increase of the total phosphorylation by 8% (NES-P). In the NES mutant, these two effects sum up linearly. The effects of the export accelerations on the nuclear phosphorylation level are different: While in the NES-U pathway the higher total phosphorylation also increases the nuclear signal, the additional export mechanism in the NES-P system counteracts the nuclear accumulation of the phosphorylated protein. In the NES mutant both effects lead to a slightly decreased nuclear concentration of phosphorylated Stat1.

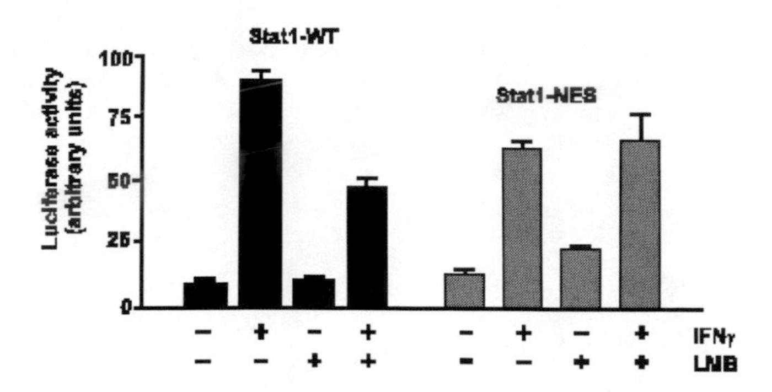

Figure 15.6: The additional export of the pY-Stat1 in the Stat1-NES mutant decreases the transcriptional response by ≈ 20%. Inhibition of the nuclear export by LMB reduces reporter gene transcription by ≈ 50% in the wild-type pathway, while the NES mutant shows no effect. Stat1 wild-type or the NES mutant were co-transfected with an IFN-γ-inducible reporter plasmid and a constitutively expressed β-galactosidase reporter gene. Cells were stimulated for 6h with IFN-γ in the presence or absence of LMB. Depicted are the mean and standard deviation from six independent experiments. (Experimental data: T. Meyer, FMP Berlin)

Export Block is Dispensable for Transcriptional Activation

Despite the lack of nuclear accumulation of the Stat1-NES mutant after cytokine stimulation, the model simulations predict that the nuclear export of the phosphorylated molecule lowers the relative nuclear phosphorylation signal in this mutant only by 20% when compared with the wild-type pathway. Interestingly, we found in subsequent experimental studies, that the mutant in fact still drives the expression of a cytokine-inducible reporter gene significantly and the reduction in luciferase activity compared to the wild-type was also only about 20% (compare Figure 15.6). As has been previously shown by the group of U. Vinkemeier, blocking nuclear export by LMB decreases the rephosphorylation rate and therefore lowers the transcriptional activity of the wild-type protein. Remarkably, for the Stat1-NES mutant no significant change of reporter gene activity was detected in the presence of LMB. This finding can be rationalised by the opposing consequences of LMB treatment on the nuclear concentration of the phospho-dimer in this phenotype. On the one hand, LMB blocks the export of the phospho-dimer; on the other hand, it also decreases the export rate of the unphosphorylated molecule and thereby retards its re-phosphorylation and re-import as phospho-protein.

The results of the numerical model simulations and of the reporter gene experiments suggest that the block of nuclear export of the phospho-protein in the wild-type pathway is dispensable for a transcriptional response, when compared to the alternative pathway design of the NES mutant discussed here. Even with a strong export of pY-Stat1 the system attains a moderate nuclear fraction of phospho-Stat1 which is sufficient for gene activation.

16

Discussion of Part III

The investigation of the different molecular mechanisms regulating signal transduction systems such as Jak/Stat1 has mainly concentrated on the activation and deactivation of the signalling proteins, for example by phosphorylation and dephosphorylation and on the modulation of DNA binding and transcriptional coactivation of the involved transcription factors. The modulatory potential of the dynamic redistribution of signalling molecules and transcriptional activators exerted by mechanisms such as nucleo-cytoplasmic shuttling and other transport processes has gained relatively little attention.

In the last part of the thesis the theoretical predictions made in part I and II regarding the distinct role of the transport processes of unphosphorylated Stat1 have been examined using experimental data of different Stat1 transport mutants. This experimental data indeed verified the previous theoretical results: The specific transport processes of active and inactive Stat1 between the nucleus and the cytoplasm are able to modulate the phosphorylation level of Stat1 and thus the system response of the signalling pathway.

Transport Processes Modulate System Response

The analysis of the behaviour of the two NLS import mutants showed that upregulation of the import rate decreases the system response. The simulation of this mutant phenotype revealed that this effect is caused by shifting the distribution of the unphosphorylated protein from the cytoplasmic to the nuclear compartment and thus constricting the access of Stat1 to the receptor-bound Jak kinase. Further analysis by numerical model simulations disclosed that the change of the phosphorylation signal is solely caused by the accelerated import of the unphosphorylated protein (k_{13}). As predicted in the control analysis in Chapter 5.1 the additional increase of import rate of the phosphorylated molecule (k_8) has no observable effect on the response of the signalling pathway.

Regulation by Export Acceleration

While the mutations in the Stat1-NLS proteins only change the kinetics of the wild-type pathway by accelerating the two nuclear import pathways of active and inactive Stat1, the NES mutation additionally disrupts the network design of the wild-type system by introducing a nuclear export mechanism for the phosphorylated molecule. The analytical calculations using the Stat core model show that this change of the network structure decreases the maximal reachable nuclear phosphorylation level to 50% (under the assumption that the phosphorylated molecule shuttles in and out of the nucleus with the same transport rate). Furthermore, the additional export increases the average dephosphorylation time for a Stat1 molecule and thus prolongates the time to deactivate the system. The numerical simulation of the NES mutant and comparison to the hypothetical mutant pathways NES-U and NES-P revealed that the experimentally observed increase of the total phosphorylation level is caused by two different effects of the NES mutation. The nuclear export of the phosphorylated protein leads to a cytoplasmic accumulation of pY-Stat1 and decreases the effective dephosphorylation by protecting the molecules from the stronger nuclear phosphatase activity. Furthermore, the NES signal leads to an increased export of the unphosphorylated molecule and facilitates its rephosphorylation in the cytoplasm. The model predicts that these two effects increase Stat phosphorylation in approximately equal shares. Interestingly, it has been recently found in the group of U. VINKEMEIER, that phosphorylation of Stat1 at serine 727 accelerates the export of the unphosphorylated molecule and increases phosphorylation [Lodige et al., 2005]. Stat1 serine phosphorylation is known to maximise target gene expression [Wen et al., 1995, Zhu et al., 1997, Goh et al., 1999, Kovarik et al., 2001] and occurs independently of the IFN-γ induced tyrosine phosphorylation [Decker and Kovarik, 2000]. These experimental results and the theoretical model analysis presented here suggest that the cell actively modulates the export of the inactive Stat1 by serine phosphorylation to increase the transcriptional response of the pathway by increased rephosphorylation of the Stat molecules.

Disruption of Network Structure Does not Inhibit Transcriptional Response

The NES mutant also revealed that the export block of the phosphorylated molecule found in the wild-type network design is dispensable for target gene expression. The additional increased phosphorylation level caused by the accelerated export and subsequent increased rephosphorylation partly cancelled the effect of the nuclear re-export of the active Stat1 molecules. Thus, the reporter gene expression was only diminished by 20%. Simulation of the hypothetical NES-P mutant, where the accelerated export of the inactive molecules did not occur showed that the disruption of the network design by removing the export block of pY-Stat1 would decrease the nuclear concentration of phosphorylated Stat1 from 25% to 17%. This nuclear phosphorylation level might still be enough to drive target gene expression. Taken together, this results reveal that the unidirectional transport of the phosphorylated molecule caused by the export block of the pY-Stat dimer increases the efficiency of the pathway, but is not essential for the activation of gene expression. As the experimental data of the NES mutant shows, the decreased nuclear phosphorylation level in a disrupted Stat1 signalling pathway with nuclear shuttling of the active molecules can be partly compensated by increased export of the inactive protein. Possibly the blocked export of the active molecules is needed to deactivate the system faster, although the differences in the simulated deactivation dynamics between NES mutant and wild-type were only marginal. Further experimentation with time-resolved measurements of the pathway deactivation are needed to test this hypothesis.

Part IV

Summary and Outlook

Summary

The Jak/Stat1 pathway is a prototypical example of a direct signal transduction pathway which transduces signals from the membrane receptor to the target genes in the nucleus without many intermediate steps and further signal processing. In this work we aimed to obtain a comprehensive understanding of Stat1 signalling and investigated the quantitative and qualitative aspects of Jak/Stat1 signal transduction. As presented in the three parts of this thesis, we approached this problem from different directions: In part I, a detailed model of the Jak/Stat1 pathway was constructed, using an iterative process of quantitative measurements of the specific pathway processes, parameter estimation, model refinement, and further experiments. The accuracy of the detailed mathematical model and of the corresponding reference parameter set was verified using independent experimental data. With the detailed mathematical description of the Stat1 pathway, we investigated the kinetic design of the network, carried out a numerical control analysis and calculated control coefficients for all major Stat pools. Thereby, we identified network processes controlling the amplitude and the duration of the system response. These analyses revealed an important regulatory potential of DNA binding, the nuclear dephosphorylation, and the nucleo-cytoplasmic transport processes of the unphosphorylated Stat1 molecule.

In part II, we derived a simplified core model of Stat signalling from the detailed pathway description. This core model allows for analytical treatment and enabled us to investigate general properties of the cyclic signalling network. We derived a relation between the steady state concentration of the different subcellular Stat pools and the lifetimes of these Stat fractions. The linear description of the Stat core model was extended to general signal transduction networks, yielding a presentation of these signalling pathways as networks of linear state transitions. We derived universal properties of such networks and a specific relation between the distribution of control among the network transition processes and the steady state occupancy of the different network states. The result of this general analysis allowed us to extend the numerical control analysis of the non-linear model presented in part I and to derive general conclusion about the control of the pathway steps over the system response solely from the network design. We could show that the process with the largest control over the nuclear concentration of active Stat1 is always the nuclear dephosphorylation reaction, independent of the rate parameters of the network processes. Furthermore, we revealed how the shuttling of the inactive transcription factor changes the control of the activating phosphorylation reaction over the nuclear pools of phosphorylated and unphosphorylated Stat1. On the one hand, the control of the Jak kinase over the nuclear concentration of active Stat1 is increased, leading to higher sensitivity of the system to changes in the extracellular stimulus. On the other hand, the control of the kinase over the nuclear pool of inactive Stat1 is decreased, leading to reduced concentration changes of unphosphorylated Stat1 during cytokine stimulation of the pathway. This reduced sensitivity of the inactive Stat1 molecules to cytokine signals might help the cell to decouple the two nuclear Stat pools and their independent transcriptional targets.

In the last part of this work we investigated the prior made theoretical predictions about the important regulatory role of the nucleo-cytoplasmic transport processes in more detail. Using different specific Stat1 mutants, the nucleo-cytoplasmic import and export rates of the phosphorylated and unphosphorylated molecules were modified and the effect of these mutations on the phosphorylation response of the pathway was investigated. Additionally, the NES mutant allowed us to study the effect of a disruption of the wild-type network structure by introducing a nuclear export pathway for phosphorylated Stat1. The experimental data of the different transport mutants verified the prior made predictions that the transport processes of the inactive Stat protein indeed are able to regulate the cytoplasmic and nuclear phosphorylation level.

Outlook

Modelling Negative Regulatory Mechanisms

The work presented here may be extended in several directions. The detailed model of the Jak/Stat1 pathway models deactivation of the system phenomenologically by a simple exponential decay of the active receptor/Jak complex. This approach proved to be sufficient to reproduce the experimentally observed system dynamics. Additionally it allowed for a simple modelling of varying stimuli patterns with different strengths and durations, enabling us to investigate the system dynamics for different stimulation modes (compare Section 5.1). However, this phenomenological approach ignores the vast biological knowledge about the molecular mechanisms controlling the deactivation of Jak/Stat1 signalling, which could be used to model the deactivation process in more detail. For example the constitutively expressed SHP phosphatases limit further Stat1 phosphorylation by tyrosine dephosphorylation of the receptor tails or Jak kinase sites [David et al., 1995, Hilton, 1999, Yasukawa et al., 2000]. The SOCS family proteins bind to receptor sites and/or Jak catalytic sites and block further Stat1 activation. Interestingly, the expression of SOCS genes is induced in response to stimulation by various cytokines via different Stat pathways [Larsen and Ropke, 2002]. Including these negative feedback mechanisms into the model could allow to study the cross-talk inhibition between the various opposing cytokine responsive Stat pathways [Greenhalgh and Hilton, 2001]. Furthermore there are a multitude of SOCS proteins with diverse functions in different cell types, opening up a wide field for investigating the various negative feedback mechanisms in Stat signalling.

Moreover, the regulation of the activity of the nuclear Stat phosphatase through the protein kinase PKR was discovered recently [Wang et al., 2006]. This negative regulatory mechanism could be included in future modelling approaches as well.

Modelling Transcriptional Activation

Based on the presented work, the model can also be extended to include a more detailed description of the transcriptional activation of the Stat target genes by the phosphorylated transcription factor. This approach could include recent experimental findings about the interaction of the Stat proteins with the transcriptional machinery [Lerner et al., 2003].

In Section 15.3 the question came up how the concentration of phosphorylated Stat1 in the nucleus determines the rate of target gene expression. Based on the artificial reporter gene data presented there, further measurements of endogenous Stat1 target gene expression (for example using the RT-PCR measurement technique) could help to elucidate the interrelation between nuclear Stat activity and target gene induction. Modelling the transcriptional activation process in more detail could include the binding of two Stat1-dimers to tandem GAS sites in certain promoters. This Stat tetramerisation process is believed to maximise transcriptional stimulation [Seidel et al., 1995, Vinkemeier et al., 1996, Horvath, 2000].

Signal Transduction as a Sequence of State Transitions

In part II of this thesis, we approximated the dynamics of the Stat signalling network by a linear differential equation system and modelled the signal transduction process as a network of state transitions. It would be interesting to see whether this method can be extended to more complex signalling pathways and whether further general conclusions about cellular signalling can be derived using this approach.

Bibliography

R.P. Andrews, M.B. Ericksen, C.M. Cunningham, M.O. Daines, and G.K. Hershey. Analysis of the life cycle of stat6. Continuous cycling of STAT6 is required for IL-4 signaling. *J Biol Chem*, 277(39):36563–9, 2002.

A.R. Asthagiri and D.A. Lauffenburger. A computational study of feedback effects on signal dynamics in a mitogen-activated protein kinase (MAPK) pathway model. *Biotechnol Prog*, 17(2):227–39, 2001.

A.R. Asthagiri, C.A. Reinhart, A.F. Horwitz, and D.A. Lauffenburger. The role of transient ERK2 signals in fibronectin- and insulin-mediated DNA synthesis. *J Cell Sci*, 113 Pt 24:4499–510, 2000.

D. Axelrod, D.E. Koppel, J. Schlessinger, E. Elson, and W.W. Webb. Mobility measurement by analysis of fluorescence photobleaching recovery kinetics. *Biophys J*, 16(9):1055–69, 1976.

G. Banninger and N.C. Reich. STAT2 nuclear trafficking. *J Biol Chem*, 279(38): 39199–206, 2004.

S. Becker, B. Groner, and C. W. Muller. Three-dimensional structure of the Stat3beta homodimer bound to DNA. *Nature*, 394(6689):145–51, 1998.

A. Begitt, T. Meyer, M. van Rossum, and U. Vinkemeier. Nucleocytoplasmic translocation of Stat1 is regulated by a leucine-rich export signal in the coiled-coil domain. *Proc Natl Acad Sci U S A*, 97(19):10418–23, 2000.

U.S. Bhalla and R. Iyengar. Emergent properties of networks of biological signaling pathways. *Science*, 283(5400):381–7, 1999.

U. Boehm, T. Klamp, M. Groot, and J. C. Howard. Cellular responses to interferon-gamma. *Annu Rev Immunol*, 15:749–95, 1997.

J. Braunstein, S. Brutsaert, R. Olson, and C. Schindler. STATs dimerize in the absence of phosphorylation. *J Biol Chem*, 278(36):34133–40, 2003.

J. Briscoe, D. Guschin, N.C. Rogers, D. Watling, M. Muller, F. Horn, P. Heinrich, G.R. Stark, and I.M. Kerr. JAKs, STATs and signal transduction in response to the interferons and other cytokines. *Philos Trans R Soc Lond B Biol Sci*, 351 (1336):167–71, 1996.

A.H. Brivanlou and J.E. Darnell, Jr. Signal transduction and the control of gene expression. *Science*, 295(5556):813–8, 2002.

G.C. Brown and B.N. Kholodenko. Spatial gradients of cellular phospho-proteins. *FEBS Lett*, 457(3):452–4, 1999.

M. Brysha, J.G. Zhang, P. Bertolino, J.E. Corbin, W.S. Alexander, N.A. Nicola, D.J. Hilton, and R. Starr. Suppressor of cytokine signaling-1 attenuates the duration of interferon gamma signal transduction in vitro and in vivo. *J Biol Chem*, 276(25):22086–9, 2001.

G. Carrero, E. Crawford, J. Th'ng, G. de Vries, and M.J. Hendzel. Quantification of protein-protein and protein-DNA interactions in vivo, using fluorescence recovery after photobleaching. *Methods Enzymol*, 375:415–42, 2004.

M. Chatterjee-Kishore, K.L. Wright, J.P. Ting, and G.R. Stark. How Stat1 mediates constitutive gene expression: a complex of unphosphorylated Stat1 and IRF1 supports transcription of the LMP2 gene. *EMBO J*, 19(15):4111–22, 2000.

T. Chen, H. L. He, and G. M. Church. Modelling gene expression with differential equations. *Proc Pacific Symp Biocomputing '99*, pages 29–40, 1999.

X. Chen, U. Vinkemeier, Y. Zhao, D. Jeruzalmi, J. E. Darnell, Jr, and J. Kuriyan. Crystal structure of a tyrosine phosphorylated STAT-1 dimer bound to DNA. *Cell*, 93(5):827–39, 1998.

A. Cornish-Bowden. *Fundamentals of Enzyme Kinetics*. Portland Press, 1995.

J.E. Darnell, Jr. STATs and gene regulation. *Science*, 277(5332):1630–5, 1997.

J.E. Darnell, Jr, I.M. Kerr, and G.R. Stark. Jak-STAT pathways and transcriptional activation in response to IFNs and other extracellular signaling proteins. *Science*, 264(5164):1415–21, 1994.

M. David, P. M. Grimley, D. S. Finbloom, and A. C. Larner. A nuclear tyrosine phosphatase downregulates interferon-induced gene expression. *Mol Cell Biol*, 13(12):7515–21, 1993.

M. David, H.E. Chen, S. Goelz, A.C. Larner, and B.G. Neel. Differential regulation of the alpha/beta interferon-stimulated Jak/Stat pathway by the SH2 domain-containing tyrosine phosphatase SHPTP1. *Mol Cell Biol*, 15(12):7050–8, 1995.

T. Decker and P. Kovarik. Serine phosphorylation of STATs. *Oncogene*, 19(21): 2628–37, 2000.

T. Decker, P. Kovarik, and A. Meinke. GAS elements: a few nucleotides with a major impact on cytokine-induced gene expression. *J Interferon Cytokine Res*, 17(3):121–34, 1997.

R. Fagerlund, K. Melen, L. Kinnunen, and I. Julkunen. Arginine/lysine-rich nuclear localization signals mediate interactions between dimeric STATs and importin alpha 5. *J Biol Chem*, 277(33):30072–8, 2002.

J.E. Ferrell and W. Xiong. Bistability in cell signaling: How to make continuous processes discontinuous, and reversible processes irreversible. *Chaos*, 11(1):227–236, 2001.

R.S. Ginger, E.C. Dalton, W.J. Ryves, M. Fukuzawa, J.G. Williams, and A.J. Harwood. Glycogen synthase kinase-3 enhances nuclear export of a Dictyostelium STAT protein. *EMBO J*, 19(20):5483–91, 2000.

F. Giordanetto and R.T. Kroemer. A three-dimensional model of Suppressor Of Cytokine Signalling 1 (SOCS-1). *Protein Eng*, 16(2):115–24, 2003.

K. C. Goh, S. J. Haque, and B. R. Williams. p38 MAP kinase is required for STAT1 serine phosphorylation and transcriptional activation induced by interferons. *EMBO J*, 18(20):5601–8, 1999.

D.M. Gowers, G.G. Wilson, and S.E. Halford. Measurement of the contributions of 1D and 3D pathways to the translocation of a protein along DNA. *Proc Natl Acad Sci U S A*, 102(44):15883–8, 2005.

C.J. Greenhalgh and D.J. Hilton. Negative regulation of cytokine signaling. *J Leukoc Biol*, 70(3):348–56, 2001.

A.C. Greenlund, M.O. Morales, B.L. Viviano, H. Yan, J. Krolewski, and R.D. Schreiber. Stat recruitment by tyrosine-phosphorylated cytokine receptors: an ordered reversible affinity-driven process. *Immunity*, 2(6):677–87, 1995.

S.E. Halford and J.F. Marko. How do site-specific DNA-binding proteins find their targets? *Nucleic Acids Res*, 32(10):3040–52, 2004.

S. E. Hartman, P. Bertone, A. K. Nath, T. E. Royce, M. Gerstein, S. Weissman, and M. Snyder. Global changes in STAT target selection and transcription regulation upon interferon treatments. *Genes Dev*, 19(24):2953–68, 2005.

R. L. Haspel and J. E. Darnell, Jr. A nuclear protein tyrosine phosphatase is required for the inactivation of Stat1. *Proc Natl Acad Sci U S A*, 96(18):10188–93, 1999.

R.L. Haspel, M. Salditt-Georgieff, and J.E. Darnell, Jr. The rapid inactivation of nuclear tyrosine phosphorylated Stat1 depends upon a protein tyrosine phosphatase. *EMBO J*, 15(22):6262–8, 1996.

M. H. Heim, I. M. Kerr, G. R. Stark, and J. E. Darnell, Jr. Contribution of STAT SH2 groups to specific interferon signaling by the Jak-STAT pathway. *Science*, 267(5202):1347–9, 1995.

P.C. Heinrich, I. Behrmann, S. Haan, H.M. Hermanns, G. Muller-Newen, and F. Schaper. Principles of interleukin (IL)-6-type cytokine signalling and its regulation. *Biochem J*, 374(Pt 1):1–20, 2003.

R. Heinrich and T.A. Rapoport. A linear steady-state treatment of enzymatic chains. General properties, control and effector strength. *Eur J Biochem*, 42(1): 89–95, 1974.

R. Heinrich and S. Schuster. *The regulation of cellular systems.* Chapman & Hall, 1996.

R. Heinrich, B.G. Neel, and T.A. Rapoport. Mathematical models of protein kinase signal transduction. *Mol Cell*, 9(5):957–70, 2002.

D.J. Hilton. Negative regulators of cytokine signal transduction. *Cell Mol Life Sci*, 55(12):1568–77, 1999.

C.M. Horvath. STAT proteins and transcriptional responses to extracellular signals. *Trends Biochem Sci*, 25(10):496–502, 2000.

A.B. Houtsmuller and W. Vermeulen. Macromolecular dynamics in living cell nuclei revealed by fluorescence redistribution after photobleaching. *Histochem Cell Biol*, 115(1):13–21, 2001.

A.B. Houtsmuller, S. Rademakers, A.L. Nigg, D. Hoogstraten, J.H. Hoeijmakers, and W. Vermeulen. Action of DNA repair endonuclease ERCC1/XPF in living cells. *Science*, 284(5416):958–61, 1999.

C.Y. Huang and J.E. Ferrell, Jr. Ultrasensitivity in the mitogen-activated protein kinase cascade. *Proc Natl Acad Sci U S A*, 93(19):10078–83, 1996.

K. Imada and W. J. Leonard. The Jak-STAT pathway. *Mol Immunol*, 37(1-2): 1–11, 2000.

G.J. Inman, F.J. Nicolas, and C.S. Hill. Nucleocytoplasmic shuttling of Smads 2, 3, and 4 permits sensing of TGF-beta receptor activity. *Mol Cell*, 10(2):283–94, 2002.

H. Kacser and J. A. Burns. The control of flux. *Symposia of the Society for Experimental Biology*, 7:65–104, 1973.

C. G. Kalodimos, N. Biris, A. M. Bonvin, M. M. Levandoski, M. Guennuegues, R. Boelens, and R. Kaptein. Structure and flexibility adaptation in nonspecific and specific protein-DNA complexes. *Science*, 305(5682):386–9, 2004.

I.M. Kerr, A.P. Costa-Pereira, B.F. Lillemeier, and B. Strobl. Of JAKs, STATs, blind watchmakers, jeeps and trains. *FEBS Lett*, 546(1):1–5, 2003.

B.N. Kholodenko. Negative feedback and ultrasensitivity can bring about oscillations in the mitogen-activated protein kinase cascades. *Eur J Biochem*, 267(6): 1583–8, 2000.

B.N. Kholodenko, G.C. Brown, and J.B. Hoek. Diffusion control of protein phosphorylation in signal transduction pathways. *Biochem J*, 350 Pt 3:901–7, 2000.

T. K. Kim and T. Maniatis. Regulation of interferon-gamma-activated STAT1 by the ubiquitin-proteasome pathway. *Science*, 273(5282):1717–9, 1996.

E. L. King and C. Altman. A schematic method of deriving the rate laws for enzyme-catalyzed reactions. *J Phys Chem*, 60:1375–1378, 1956.

T. Kisseleva, S. Bhattacharya, J. Braunstein, and C.W. Schindler. Signaling through the JAK/STAT pathway, recent advances and future challenges. *Gene*, 285(1-2):1–24, 2002.

I. Koch, C. Segev, editor. *Methods in Neuronal Modeling: From Ions to Networks.* Bradford Book MIT Press, 1998.

A. Komeili and E.K. O'Shea. Nuclear transport and transcription. *Curr Opin Cell Biol*, 12(3):355–60, 2000.

A. Komeili and E.K. O'Shea. New perspectives on nuclear transport. *Annu Rev Genet*, 35:341–64, 2001.

P. Kovarik, M. Mangold, K. Ramsauer, H. Heidari, R. Steinborn, A. Zotter, D. E. Levy, M. Muller, and T. Decker. Specificity of signaling by STAT1 depends on SH2 and C-terminal domains that regulate Ser727 phosphorylation, differentially affecting specific target gene expression. *EMBO J*, 20(1-2):91–100, 2001.

L. Larsen and C. Ropke. Suppressors of cytokine signalling: SOCS. *APMIS*, 110 (12):833–44, 2002.

J.F. Lau and C.M. Horvath. Mechanisms of Type I interferon cell signaling and STAT-mediated transcriptional responses. *Mt Sinai J Med*, 69(3):156–68, 2002.

C.K. Lee, H.A. Bluyssen, and D.E. Levy. Regulation of interferon-alpha responsiveness by the duration of Janus kinase activity. *J Biol Chem*, 272(35):21872–7, 1997.

E. Lee, A. Salic, R. Kruger, R. Heinrich, and M.W. Kirschner. The roles of APC and Axin derived from experimental and theoretical analysis of the Wnt pathway. *PLoS Biol*, 1(1):E10, 2003.

L. Lerner, M.A. Henriksen, X. Zhang, and J.E. Darnell, Jr. STAT3-dependent enhanceosome assembly and disassembly: synergy with GR for full transcriptional increase of the alpha 2-macroglobulin gene. *Genes Dev*, 17(20):2564–77, 2003.

S. Leung, X. Li, and G. R. Stark. STATs find that hanging together can be stimulating. *Science*, 273(5276):750–1, 1996.

K. Levenberg. A method for the solution of certain problems in least squares. *Quart Appl Math*, 2:164–168, 1944.

D.E. Levy and J.E. Darnell, Jr. Stats: transcriptional control and biological impact. *Nat Rev Mol Cell Biol*, 3(9):651–62, 2002.

B.F. Lillemeier, M. Koster, and I.M. Kerr. STAT1 from the cell membrane to the DNA. *EMBO J*, 20(10):2508–17, 2001.

T. Lipniacki, P. Paszek, A.R. Brasier, B. Luxon, and M. Kimmel. Mathematical model of NF-kappaB regulatory module. *J Theor Biol*, 228(2):195–215, 2004.

J. Lippincott-Schwartz, N. Altan-Bonnet, and G.H. Patterson. Photobleaching and photoactivation: following protein dynamics in living cells. *Nat Cell Biol*, Suppl:S7–14, 2003.

L. Liu, K.M. McBride, and N.C. Reich. STAT3 nuclear import is independent of tyrosine phosphorylation and mediated by importin-alpha3. *Proc Natl Acad Sci U S A*, 102(23):8150–5, 2005.

I. Lodige, A. Marg, B. Wiesner, B. Malecova, T. Oelgeschlager, and U. Vinkemeier. Nuclear export determines the cytokine sensitivity of STAT transcription factors. *J Biol Chem*, 280(52):43087–99, 2005.

X. Mao, Z. Ren, G.N. Parker, H. Sondermann, M.A. Pastorello, W. Wang, J.S. McMurray, B. Demeler, J.E. Darnell, Jr, and X. Chen. Structural bases of unphosphorylated STAT1 association and receptor binding. *Mol Cell*, 17(6): 761–71, 2005.

A. Marg, Y. Shan, T. Meyer, T. Meissner, M. Brandenburg, and U. Vinkemeier. Nucleocytoplasmic shuttling by nucleoporins Nup153 and Nup214 and CRM1-dependent nuclear export control the subcellular distribution of latent Stat1. *J Cell Biol*, 165(6):823–33, 2004.

D. Marquardt. An alogrithm for least-squares estimation of nonlinear parameters. *SIAM J Appl Math*, 11:431–441, 1963.

C.J. Marshall. Specificity of receptor tyrosine kinase signaling: transient versus sustained extracellular signal-regulated kinase activation. *Cell*, 80(2):179–85, 1995.

R. Martone, G. Euskirchen, P. Bertone, S. Hartman, T. E. Royce, N. M. Luscombe, J. L. Rinn, F. K. Nelson, P. Miller, M. Gerstein, S. Weissman, and M. Snyder. Distribution of NF-kappaB-binding sites across human chromosome 22. *Proc Natl Acad Sci U S A*, 100(21):12247–52, 2003.

MathWorks. Optimization Toolbox User's Guide Version 2. http://www.mathworks.com/access/helpdesk/help/toolbox/optim/optim.html, 3 Apple Hill Drive, Natrick, MMA 01760-2098, USA, 2003.

K.M. McBride, C. McDonald, and N.C. Reich. Nuclear export signal located within theDNA-binding domain of the STAT1transcription factor. *EMBO J*, 19 (22):6196–206, 2000.

K.M. McBride, G. Banninger, C. McDonald, and N.C. Reich. Regulated nuclear import of the STAT1 transcription factor by direct binding of importin-alpha. *EMBO J*, 21(7):1754–63, 2002.

M.A. Meraz, J.M. White, K.C. Sheehan, E.A. Bach, S.J. Rodig, A.S. Dighe, D.H. Kaplan, J.K. Riley, A.C. Greenlund, D. Campbell, K. Carver-Moore, R.N. DuBois, R. Clark, M. Aguet, and R.D. Schreiber. Targeted disruption of the Stat1 gene in mice reveals unexpected physiologic specificity in the JAK-STAT signaling pathway. *Cell*, 84(3):431–42, 1996.

D. Metcalf, S. Mifsud, L. Di Rago, N.A. Nicola, D.J. Hilton, and W.S. Alexander. Polycystic kidneys and chronic inflammatory lesions are the delayed consequences of loss of the suppressor of cytokine signaling-1 (SOCS-1). *Proc Natl Acad Sci U S A*, 99(2):943–8, 2002.

J. T. Mettetal, D. Muzzey, J. M. Pedraza, E. M. Ozbudak, and A. van Oudenaarden. Predicting stochastic gene expression dynamics in single cells. *Proc Natl Acad Sci U S A*, 103(19):7304–9, 2006.

T. Meyer, A. Begitt, I. Lodige, M. van Rossum, and U. Vinkemeier. Constitutive and IFN-gamma-induced nuclear import of STAT1 proceed through independent pathways. *EMBO J*, 21(3):344–54, 2002a.

T. Meyer, K. Gavenis, and U. Vinkemeier. Cell type-specific and tyrosine phosphorylation-independent nuclear presence of STAT1 and STAT3. *Exp Cell Res*, 272(1):45–55, 2002b.

T. Meyer, A. Marg, P. Lemke, B. Wiesner, and U. Vinkemeier. DNA binding controls inactivation and nuclear accumulation of the transcription factor Stat1. *Genes Dev*, 17(16):1992–2005, 2003.

T. Meyer, L. Hendry, A. Begitt, S. John, and U. Vinkemeier. A single residue modulates tyrosine dephosphorylation, oligomerization, and nuclear accumulation of stat transcription factors. *J Biol Chem*, 279(18):18998–9007, 2004.

A. Miyawaki, A. Sawano, and T. Kogure. Lighting up cells: labelling proteins with fluorophores. *Nat Cell Biol*, Suppl:S1–7, 2003.

K. Mowen and M. David. Regulation of STAT1 nuclear export by Jak1. *Mol Cell Biol*, 20(19):7273–81, 2000.

K.A. Mowen, J. Tang, W. Zhu, B.T. Schurter, K. Shuai, H.R. Herschman, and M. David. Arginine methylation of STAT1 modulates IFNalpha/beta-induced transcription. *Cell*, 104(5):731–41, 2001.

F.J. Nicolas, K. De Bosscher, B. Schmierer, and C.S. Hill. Analysis of Smad nucleocytoplasmic shuttling in living cells. *J Cell Sci*, 117(Pt 18):4113–25, 2004.

H. Okamura, C. Garcia-Rodriguez, H. Martinson, J. Qin, D.M. Virshup, and A. Rao. A conserved docking motif for CK1 binding controls the nuclear localization of NFAT1. *Mol Cell Biol*, 24(10):4184–95, 2004.

J.A. Papin and B.O. Palsson. The JAK-STAT signaling network in the human B-cell: an extreme signaling pathway analysis. *Biophys J*, 87(1):37–46, 2004.

J. M. Pedraza and A. van Oudenaarden. Noise propagation in gene networks. *Science*, 307(5717):1965–9, 2005.

R.D. Phair and T. Misteli. High mobility of proteins in the mammalian cell nucleus. *Nature*, 404(6778):604–9, 2000.

R.D. Phair and T. Misteli. Kinetic modelling approaches to in vivo imaging. *Nat Rev Mol Cell Biol*, 2(12):898–907, 2001.

A.L. Pranada, S. Metz, A. Herrmann, P.C. Heinrich, and G. Muller-Newen. Real time analysis of STAT3 nucleocytoplasmic shuttling. *J Biol Chem*, 279(15): 15114–23, 2004.

W. P. Press, S. A. Teukolsky, W. T. Vetterling, and B. P. Flannery. *Numerical Recipes in C++*. Cambridge University Press, 2002.

N. Rosenfeld, J. W. Young, U. Alon, P. S. Swain, and M. B. Elowitz. Gene regulation at the single-cell level. *Science*, 307(5717):1962–5, 2005.

C. Schindler, K. Shuai, V. R. Prezioso, and J. E. Darnell, Jr. Interferon-dependent tyrosine phosphorylation of a latent cytoplasmic transcription factor. *Science*, 257(5071):809–13, 1992.

U. Schindler, P. Wu, M. Rothe, M. Brasseur, and S. L. McKnight. Components of a Stat recognition code: evidence for two layers of molecular selectivity. *Immunity*, 2(6):689–97, 1995.

H.M. Seidel, L.H. Milocco, P. Lamb, J.E. Darnell, Jr, R.B. Stein, and J. Rosen. Spacing of palindromic half sites as a determinant of selective STAT (signal transducers and activators of transcription) DNA binding and transcriptional activity. *Proc Natl Acad Sci U S A*, 92(7):3041–5, 1995.

T. Sekimoto, N. Imamoto, K. Nakajima, T. Hirano, and Y. Yoneda. Extracellular signal-dependent nuclear import of Stat1 is mediated by nuclear pore-targeting complex formation with NPI-1, but not Rch1. *EMBO J*, 16(23):7067–77, 1997.

K. Shuai, C. Schindler, V. R. Prezioso, and J. E. Darnell, Jr. Activation of transcription by IFN-gamma: tyrosine phosphorylation of a 91-kD DNA binding protein. *Science*, 258(5089):1808–12, 1992.

K. Shuai, A. Ziemiecki, A. F. Wilks, A. G. Harpur, H. B. Sadowski, M. Z. Gilman, and J. E. Darnell. Polypeptide signalling to the nucleus through tyrosine phosphorylation of Jak and Stat proteins. *Nature*, 366(6455):580–3, 1993.

K. Shuai, C. M. Horvath, L. H. Huang, S. A. Qureshi, D. Cowburn, and J. E. Darnell, Jr. Interferon activation of the transcription factor Stat91 involves dimerization through SH2-phosphotyrosyl peptide interactions. *Cell*, 76(5):821–8, 1994.

I. Swameye, T.G. Muller, J. Timmer, O. Sandra, and U. Klingmuller. Identification of nucleocytoplasmic cycling as a remote sensor in cellular signaling by databased modeling. *Proc Natl Acad Sci U S A*, 100(3):1028–33, 2003.

J. ten Hoeve, M. de Jesus Ibarra-Sanchez, Y. Fu, W. Zhu, M. Tremblay, M. David, and K. Shuai. Identification of a nuclear Stat1 protein tyrosine phosphatase. *Mol Cell Biol*, 22(16):5662–8, 2002.

T. Tenev, S.A. Bohmer, R. Kaufmann, S. Frese, T. Bittorf, T. Beckers, and F.D. Bohmer. Perinuclear localization of the protein-tyrosine phosphatase SHP-1 and inhibition of epidermal growth factor-stimulated STAT1/3 activation in A431 cells. *Eur J Cell Biol*, 79(4):261–71, 2000.

S.J. Vayttaden, S.M. Ajay, and U.S. Bhalla. A spectrum of models of signaling pathways. *Chembiochem*, 5(10):1365–74, 2004.

U. Vinkemeier, S.L. Cohen, I. Moarefi, B.T. Chait, J. Kuriyan, and J.E. Darnell, Jr. DNA binding of in vitro activated Stat1 alpha, Stat1 beta and truncated Stat1: interaction between NH2-terminal domains stabilizes binding of two dimers to tandem DNA sites. *EMBO J*, 15(20):5616–26, 1996.

S. Wang, J.F. Raven, D. Baltzis, S. Kazemi, D.V. Brunet, M. Hatzoglou, M.L. Tremblay, and A.E. Koromilas. The catalytic activity of the eukaryotic initiation factor-2alpha kinase PKR is required to negatively regulate Stat1 and Stat3 via activation of the T-cell protein-tyrosine phosphatase. *J Biol Chem*, 281(14): 9439–49, 2006.

Z. Wen, Z. Zhong, and J. E. Darnell, Jr. Maximal activation of transcription by Stat1 and Stat3 requires both tyrosine and serine phosphorylation. *Cell*, 82(2): 241–50, 1995.

T. R. Wu, Y. K. Hong, X. D. Wang, M. Y. Ling, A. M. Dragoi, A. S. Chung, A. G. Campbell, Z. Y. Han, G. S. Feng, and Y. E. Chin. SHP-2 is a dual-specificity phosphatase involved in Stat1 dephosphorylation at both tyrosine and serine residues in nuclei. *J Biol Chem*, 277(49):47572–80, 2002.

L. Xu and J. Massague. Nucleocytoplasmic shuttling of signal transducers. *Nat Rev Mol Cell Biol*, 5(3):209–19, 2004.

E. Yang, M. A. Henriksen, O. Schaefer, N. Zakharova, and J. E. Darnell, Jr. Dissociation time from DNA determines transcriptional function in a STAT1 linker mutant. *J Biol Chem*, 277(16):13455–62, 2002.

J. Yang, M. Chatterjee-Kishore, S.M. Staugaitis, H. Nguyen, K. Schlessinger, D.E. Levy, and G.R. Stark. Novel roles of unphosphorylated STAT3 in oncogenesis and transcriptional regulation. *Cancer Res*, 65(3):939–47, 2005.

H. Yasukawa, A. Sasaki, and A. Yoshimura. Negative regulation of cytokine signaling pathways. *Annu Rev Immunol*, 18:143–64, 2000.

M. You, D.H. Yu, and G.S. Feng. Shp-2 tyrosine phosphatase functions as a negative regulator of the interferon-stimulated Jak/STAT pathway. *Mol Cell Biol*, 19(3):2416–24, 1999.

R. Zeng, Y. Aoki, M. Yoshida, K. Arai, and S. Watanabe. Stat5B shuttles between cytoplasm and nucleus in a cytokine-dependent and -independent manner. *J Immunol*, 168(9):4567–75, 2002.

J.G. Zhang, D. Metcalf, S. Rakar, M. Asimakis, C.J. Greenhalgh, T.A. Willson, R. Starr, S.E. Nicholson, W. Carter, W.S. Alexander, D.J. Hilton, and N.A. Nicola. The SOCS box of suppressor of cytokine signaling-1 is important for inhibition of cytokine action in vivo. *Proc Natl Acad Sci U S A*, 98(23):13261–5, 2001.

W. Zhu, T. Mustelin, and M. David. Arginine methylation of STAT1 regulates its dephosphorylation by T cell protein tyrosine phosphatase. *J Biol Chem*, 277 (39):35787–90, 2002.

X. Zhu, Z. Wen, L. Z. Xu, and J. E. Darnell, Jr. Stat1 serine phosphorylation occurs independently of tyrosine phosphorylation and requires an activated Jak2 kinase. *Mol Cell Biol*, 17(11):6618–23, 1997.

Z. Zi, K.H. Cho, M.H. Sung, X. Xia, J. Zheng, and Z. Sun. In silico identification of the key components and steps in IFN-gamma induced JAK-STAT signaling pathway. *FEBS Lett*, 579(5):1101–8, 2005.

Part V

Appendix

A

Modelling Stat1 Dimerisation

The cytoplasmic phosphorylation (k_5), dimerization (k_6), dimer breakup (k_{-6}) and dephosphorylation (k_7) reactions are described by the following differential equation system:

$$\frac{d}{dt} Y_c^{p,m} = k_5 Y_r - 2k_6 Y_c^{p,m^2} + 2k_{-6} Y_c^{p,d} - k_7 Y_c^{p,m} \quad \text{(A.1)}$$

$$\frac{d}{dt} Y_c^{p,d} = k_6 Y_c^{p,m^2} - (k_{-6} + k_7 + k_8) Y_c^{p,d} \quad \text{(A.2)}$$

where $Y_c^{p,m}$ and $Y_c^{p,d}$ denote the cytoplasmic Stat1 monomer and dimer concentrations, respectively and k_8 refers to the nuclear import of pY-Stat1 dimers.

We introduce as a new variable the sum of the phosphorylated Stat concentrations in the cytoplasm, measured in monomer units:

$$Y_c^p = Y_c^{p,m} + 2Y_c^{p,d} \quad \text{(A.3)}$$

Adding equations (A.1) and (A.2) according to (A.3) yields the differential equation for Y_c^p. Assuming a rapid and high-affinity dimerization reaction [Greenlund et al., 1995] implies that $Y_c^{p,m}$ is much smaller than $Y_c^{p,d}$, so we can neglect $Y_c^{p,m}$ and approximate $Y_c^p \approx 2Y_c^{p,d}$. The differential equation for Y_c^p than simplifies to

$$\frac{d}{dt} Y_c^p = k_5 Y_r - (k_7 + k_8) Y_c^p \quad \text{(A.4)}$$

B

Abbrevations and Symbols

Abbrevations

AU arbitrary units

Crm1 chromosomal region maintenance/exportin 1, export receptor

cyt cytoplasmic

ΔNLS Stat1 transport mutant protein, see NLS

DNAminus Stat1 DNA binding mutant protein

DNAplus Stat1 DNA binding mutant protein

dsNLS dimer-specific NLS

FRAP fluorescence recovery after photobleaching

GAS gamma activated sites/sequences

GFP green fluorescent protein

HeLa cell type expressing endogenous Stat1

IFN see IFN-γ

IFN-γ interferon-γ, cytokine

Jak Janus tyrosine kinase

LMB leptomycin B, inhibitor of Crm1 dependent export

MC Monte Carlo

MCA metabolic control analysis

MG132 proteasome inhibitor

min minute

NES nuclear export signal

NF-κB nuc. factor of kappa light chain gene enhancer, transcription factor

NFAT nuclear factor of activated T-cells, transcription factor

nM nanomolar

NLS nuclear localisation signal

NPC nuclear pore complex

nuc nuclear

PIAS protein inhibitor of activated Stat, inhibitors of Jak/Stat signalling

phospho-Stat see pY-Stat

pY-Stat Stat phosphorylated at the tyrosine residue

ROI region of interest

SH2 Src homology 2, protein domain

Stat signal transducer and activator of transcription, transcription factor

Stat1-GFP fusion protein of Stat1 and GFP

SOCS suppressors of cytokine signaling, inhibitors of Jak/Stat signalling

TF transcription factor

U3A Stat1-negative cell type

WT wild-type

Y-Stat Stat not phosphorylated at the tyrosine residue

Symbols

$[\alpha]$	curvature matrix
α	response amplitude
C	estimated covariance matrix
$C_j^{Y_i}$	concentration control coefficient of parameter j over concentration Y_i
C_j^{Tot}	total network control of parameter j
D_S	Stat1 diffusion coefficient
$\mathcal{D}_0$	initial simulated dataset
$\mathcal{D}_e$	experimentally measured dataset
$\mathcal{D}_i$	simulated dataset
$\vec{F}(\vec{p})$	residual vector
$\vec{F}_i(\vec{p})$	i-th component of $\vec{F}(\vec{p})$
$H(\vec{p})$	Hessian matrix
$H_i(\vec{p})$	Hessian of $\vec{F}_i(\vec{p})$
I_0	fluorescence intensity at t=0
I_∞	fluorescence intensity at equilibrium
I_p	integrated response of nuclear pY-Stat1
$J(\vec{p})$	Jacobian matrix
J_{SS}	steady-state flux
k_i	rate constant (forward direction)
k_{-i}	rate constant (backward direction)
N_S	number of Stat molecules in the cell
$\phi(t, \vec{p})$	fluorescence intensity model function
p_i	parameter
$\vec{p}$	parameter vector
$\vec{p}_0$	boot-strap parameter vector
$\vec{p}_{est}$	estimated parameter vector
$\vec{p}_{true}$	true parameter vector
P_n	nuclear pY-Stat1
ρ	ratio of cytoplasmic and nuclear volumes
R	concentration of unphosphorylated receptor/Jak complex
R_0	initial amount of active receptor/Jak complex, stimulus strength
R^p	concentration of phosphorylated receptor/Jak complex
$R_{i,j}$	correlation coefficient between parameter i and j
$\mathcal{S}_i$	King-Altman sub-path product
S_i	concentration of node i
$\overline{S}_i$	steady state concentration of node i
$\mathcal{S}_T$	sum off all allowed King-Altman sub-path products in a network
σ_i	variance of datapoint i
$\sigma_{p,n}$	response duration

τ_C	cycle time
$\tau_{p,n}$	average nuclear residence time of pY-Stat1
$t_{1/2}$	half-time of pancellular steady-state distribution
$T_{0.5}$	stimulus half-life
v_i	reaction rate
V	volume
V_{cyt}	cytoplasmic volume
V_{nuc}	nuclear volume
Y	concentration of free Stat1
Y_c	concentration of unphosphorylated cytoplasmic Stat1
Y_c^p	concentration of phosphorylated cytoplasmic Stat1
$Y_c^{p,d}$	concentration of phosphorylated cytoplasmic Stat1 dimers
$Y_c^{p,m}$	concentration of phosphorylated cytoplasmic Stat1 monomers
Y_n	concentration of unphosphorylated nuclear Stat1
Y_n^p	concentration of phosphorylated nuclear Stat1
Y_r	Stat1 bound to the receptor/Jak complex
Z	concentration of DNA bound Stat1
Z_s	phosphorylated Stat1 bound to GAS sites
Z_u	phosphorylated Stat1 bound unspecific DNA sites

Danksagung

An erster Stelle möchte ich meinem Betreuer, Prof. Thomas Höfer für seine fachlich und menschlich erstklassige Betreuung und für die Möglichkeit in seiner Arbeitsgruppe zu arbeiten, danken. Thomas hat mir bei allen Problemen Mut gemacht und immer wieder neue Lösungsansätze aufgezeigt.

Mein Dank gilt auch besonders Dr. Uwe Vinkemeier und Dr. Thomas Meyer von der von Dr. Vinkemeier geleiteten Forschungsgruppe Zelluläre Signalverarbeitung am Forschungsinstitut Molekulare Pharmakologie in Berlin-Buch. Ohne die zahlreichen wissenschaftlichen Diskussionen und Ihr kritisches Feedback wäre diese Arbeit nicht möglich gewesen. Herausheben möchte ich in diesem Zusammenhang auch die zahllosen experimentellen Messungen, die von Dr. Meyer und anderen Mitgliedern der Arbeitsgruppe auf unseren Wunsch hin durchgeführt wurden und auf denen viele der hier vorgestellten Ergebnisse basieren. Ich möchte auch Herrn Dr. Burkhard Wiesner für die Hilfe bei der Durchführung der FRAP-Messungen danken.

Ich bedanke mich auch bei dem Leiter des Graduiertenkollegs 268, Prof. Reinhardt Heinrich, für die finanzielle Unterstützung und für die zahllosen Möglichkeiten an verschiedenen Konferenzen und Workshops teilzunehmen.

Besonders danke ich Dr. Nils Blüthgen für die wissenschaftlichen Diskussionen und auch sonst angenehmen Unterhaltungen, die wir regelmäßig beim gemeinsamen Mittagessen in der Bauernmensa führten. Außerdem basiert das Erscheinungsbild dieser Arbeit auf Nils' Latex-Style und er hat netterweise mein Manuskript korrekturgelesen.

Ich möchte mich auch allen Mitgliedern der Arbeitsgruppen Höfer und Heinrich bedanken, die ein so angenehmes und nettes Arbeitsklima geschaffen haben. Antonio Politi danke ich für die Einführung in das Gebiet der Parameter-Schätzung. Besonderer Dank gilt hier auch Dorothea Busse und Jana Schütze für das Korrigieren dieser Arbeit.

Bedanken möchte ich mich auch nochmals bei Prof. Hanspeter Herzel, der 2002/2003 meine Diplomarbeit betreut hat. Er führte mich damals in das mir unbekannte Gebiet der Systembiologie ein und ebnete somit den Weg zu dieser Dissertationsarbeit.

Ich bedanke mich bei unseren Systemadministratoren Dr. Robert Arndt und Andreas Lehmann dafür, dass unser Rechner-Netzwerk immer lief und Probleme immer in kürzester Zeit behoben wurden. Außerdem danke ich auch unserer Instituts-Sekretärin Silvia Zielsdorf, die bei administrativen Problemen immer Rat wusste und alles hervorragend organisiert hat.

Schließlich möchte ich meinen Eltern Antonia und Paul und allen meinen Freunden für Ihre Unterstützung danken.

Selbständigkeitserklärung

Ich versichere hiermit, die vorliegende Arbeit selbständig und ausschließlich unter Verwendung der angegebenen Mittel und ohne unerlaubte Hilfen angefertigt zu haben.

Berlin, den 15. September 2006

Stephan Beirer